Nazeran Idrees, PhD

AF880915

Essentials of Graph Theory

First Edition

ISBN: 978-1-64953-973-1

ABSOLUTE AUTHOR PUBLISHING HOUSE

Preface

Graph theory, a branch of discrete mathematics, has emerged as an indispensable framework for understanding the intricate connections that underlie complex systems in diverse disciplines. As we stand on the precipice of the information age, where networks and relationships shape our interactions and decisions, the significance of graph theory has never been more pronounced.

This book, "Graph Theory Essentials: A Journey through Fundamental Concepts," is born out of a passion for demystifying the intricacies of graph theory and making its principles accessible to a broad audience. Whether you are a student taking your first steps into the realm of graphs or a seasoned professional seeking a refresher, our goal is to provide you with a clear and comprehensive exploration of the core concepts that define graph theory.

In crafting this book, we have striven to strike a delicate balance between theoretical rigor and practical applicability. Theoretical foundations are laid out with precision, allowing readers to grasp the essence of graph theory, while real-world applications showcase the versatility and relevance of these abstract notions. The aim is to equip you with not only the knowledge to navigate the intricacies of graph theory but also the tools to apply these principles in practical problem-solving scenarios.

Each chapter unfolds a new layer of understanding, building upon the concepts introduced earlier. The narrative is enriched with illustrative examples, challenging exercises, and insightful anecdotes to engage and stimulate your curiosity. We invite you to delve into the pages of this book with an open mind, ready to explore the beauty and utility of graph theory.

Graph theory, with its roots dating back to the eighteenth century, has evolved into a discipline that transcends academic boundaries. From computer science to biology, sociology to logistics, graphs offer a universal language to articulate relationships and patterns. The applications are boundless, ranging from optimizing routes in transportation networks to unraveling the intricacies of social connections on digital platforms.

As we embark on this journey together, I encourage you to embrace the spirit of inquiry and discovery. Graph theory is not merely a collection of theorems and definitions but a living, breathing language that speaks to the interconnectedness of the world around us. Whether you find yourself intrigued by the elegance of mathematical structures or fascinated by the practical applications in our technologically driven society, there is something in these pages for everyone.

So, with enthusiasm and curiosity, let us venture into the realm of nodes and edges, traverse the landscapes of connectivity and cycles, and unlock the doors to a world where graph theory is not just a subject of study but a powerful lens through which we can perceive and comprehend the intricacies of our interconnected world.

Happy reading and happy exploring!

Nazeran Idrees, PhD

December 2023

Table of Contents

Chapter 1

Basic Notions of Graph Theory

History

Graph theory is a branch of mathematics that studies the relationships between nodes and edges in graphs. The history of graph theory can be traced back to the 18th century, and its development has been influenced by various mathematical, scientific, and practical applications. Here's a brief overview of the key milestones in the history of graph theory:

1. **Leonhard Euler (1759):** Euler continued to develop graph theory in his paper "Recherches sur le problème des moulins à vent" ("Investigations on the theory of the game of moulins"), published in 1759. In this work, Euler introduced the concept of a graph and established fundamental principles, such as the degree of a vertex and the concept of connected graphs.
2. **Gustav Kirchhoff (1847):** Kirchhoff, a German physicist, used graph theory to analyze electrical circuits. He introduced the concept of a graph's tree and the Kirchhoff laws, which relate the currents and voltages in a network of resistors.
3. **Arthur Cayley (1857):** Cayley, a British mathematician, made significant contributions to the enumeration of trees and the theory of permutations. He was the first to use the term "graph" in a paper titled "On the Theory of the Analytical Forms Called Trees."
4. **William Tutte (1940s):** Tutte, a British-Canadian mathematician, made important contributions during World War II to solve military communication problems. His work on graph theory, including the concept of a "Tutte matrix," became influential in the study of graph theory.
5. **Paul Erdős and Alfred Rényi (1959):** They introduced the random graph model, which has since become a fundamental concept in the study of graph theory. The

Erdős–Rényi model describes a graph with a fixed number of vertices where edges are added between pairs of vertices randomly.

6. **Graph Theory in Computer Science (20th Century):** Graph theory found extensive applications in computer science, especially in the representation and analysis of data structures. Concepts like depth-first search and breadth-first search, used for graph traversal, became crucial in algorithm design.
7. **Modern Developments (20th Century Onward):** Graph theory has continued to evolve with contributions from various mathematical disciplines, computer science, and applications in diverse fields such as social networks, biology, transportation, and optimization.

The development of graph theory has been characterized by a combination of theoretical insights and practical applications, making it a versatile and influential branch of mathematics.

Definition 1

A graph is a mathematical structure consisting of vertices (or nodes) and edges (or links) that connect pairs of vertices. The vertices and edges may have additional attributes or properties, and graphs can be used to model and analyze relationships between various entities.

Weighted Graph: A graph in which each edge has an associated numerical value or weight. Weighted graphs are used to model scenarios where the connections between vertices have specific magnitudes.

Example 1

Let us take a Graph $G = (V,E)$, where

Vertices, $V = \{h,i,j,f\}$

Edges, $E = \{\{h,i\},\{h,j\},\{i, j\},\{j,f\}\}$

Example 2

In this graph, there exists a path connecting any pair of vertices, allowing travel from one vertex to any other. For instance, one can navigate from vertex 'a' to vertex 'e' by following the path 'a-b-e.'

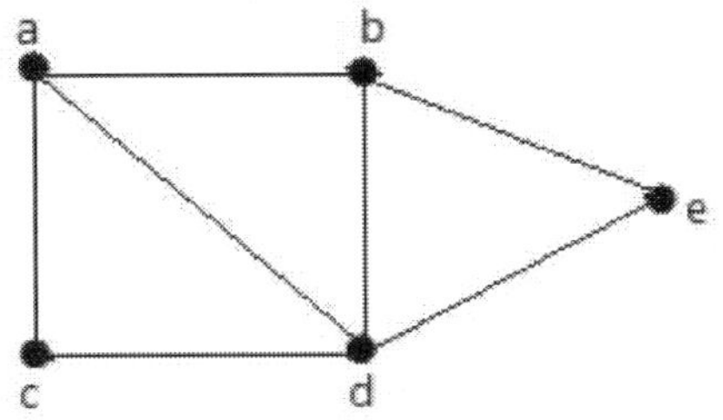

Example 3

In the graphs given below, navigating from vertex ‘a’ to vertex ‘f’ is not possible because there is no path between them, proving it to be a disconnected graph.

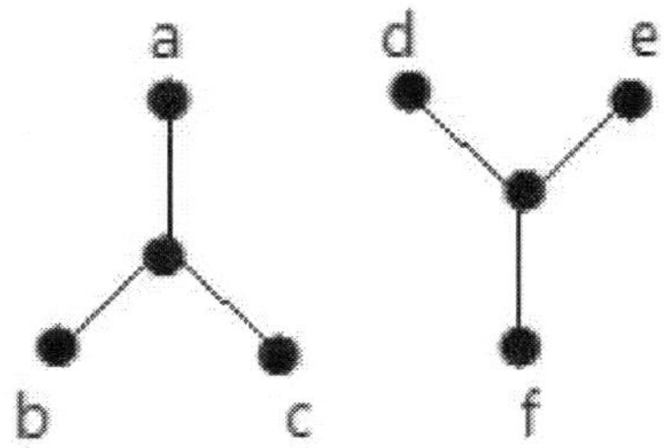

Remarks:

Two vertices, say, x and y are said to be connected ina graph if there is a path in the graph starting from x and ending at y.

Connectivity is an equivalence relation then there is a partition of V(vertices) into non-empty subsets $V1, V2, V3, \ldots, Vr,$ which are disjoint. If $u \text{ } and \text{ } v$ are connected, they lie in same class.

The sub graphs $G(V1), G(V2), G(V3), \ldots, G(Vr)$ are called connected components of G.

If G has exactly one connected component, then G is called connected otherwise it is disconnected.

Incidence Matrix of a Graph

To any graph G, there corresponds $|V| \times |E|$ matrix, denoted by $M(G) = [m_{ij}]$,

Where m_{ij} is the number of times v_i and e_j are incident.

Example 4

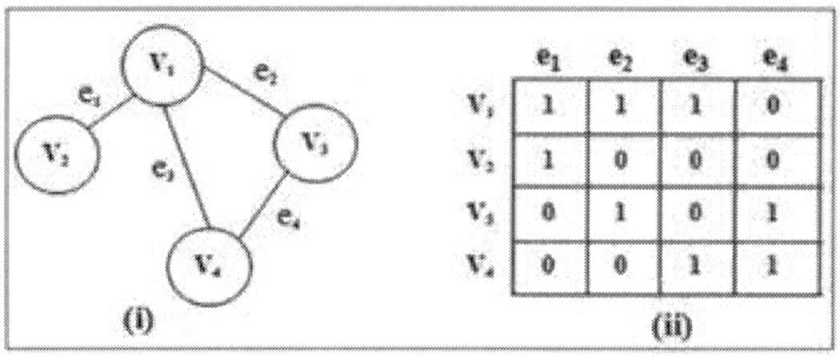

Remark 1

In general, sum of each column in incident matrix M of a graph is 2 always.

Sum of all entries of incidence matrix M is $2|E|$.

Adjacency Matrix of Graph

To any graph $G\ (V, E)$, there corresponds $|V| \times |V|$, denoted by $A(G) = [a_{ij}]$.

Where $a_{ij} = \begin{cases} 1 & if\ v_i v_j \in E(G) \\ 0 & otherwise \end{cases}$

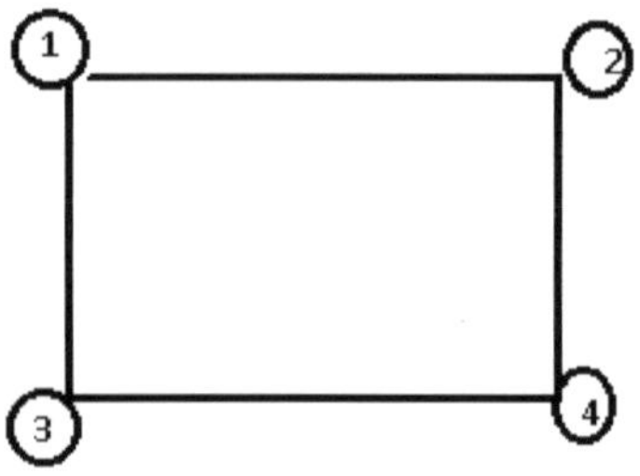

Has adjacency matrix

0	1	0	1

1	0	1	0
0	1	0	1
1	0	1	0

Note: $A(G) = [a_{ij}]$ is a symmetric matrix.

Handshaking Lemma

Let G = (V, E) be a simple graph then $\sum_{v \in V} d(v) = 2\,|E|$.

Proof:

Consider the incident matrix of the graph G. The sum of entries in a row of the matrix is precisely $d(v)$. So, $\sum_{v \in V} d(v)$ is the sum of all entries of M.

The sum of entries of each column of incidence matrix M is 2. So, the sum of all entries of M is 2 |E|.

This implies that $\sum_{v \in V} d(V) = 2\,|E|$.

Corollary 1

Let $G = (V, E)$ be a graph. Then there is even number of vertices in V with odd degree.

Proof:

Let V_1 and V_2 be the subsets of V of vertices of odd degree and even degree respectively.

By handshaking lemma,

$\sum_{v \in V1} d\,(v) + \sum_{v \in V2} d\,(v) = \sum_{v \in V} d\,(v) = 2\,|E|$.

Now, since $\sum_{v \in V2} d\,(v)$ is even; so is $\sum_{v \in V1} d(v)$.

Since $d(v)$ is odd for all $v \in V_1$. So V_1 has even number of vertices.i.e. Vertices having odd degree are even in numbers.

Definition 2

Non-trivial simple graphs must have at least one pair of vertices whose degrees are same.

Trivial simple graph is a graph which has only one vertex and no edge.

Pigeonhole Principle

The Pigeonhole Principle is a fundamental principle in combinatorics, a branch of mathematics dealing with counting and arranging objects. The principle is often stated in a simple and intuitive manner:

"If you want to distribute n items into m boxes and n > m, then at least one box must have more than one item."

In other words, if you are trying to assign more items than there are boxes, then at least one box must have more than one item. The analogy often used is that if you have more pigeons than pigeonholes, there must be at least one pigeonhole with more than one pigeon.

The Pigeonhole Principle is widely applied in various areas of mathematics as well as computer science. It helps establish the existence of repetitions, patterns, or constraints in a given context.

Theorem 1: Consider a non-trivial graph G must contain at least two nodes with equal degrees.

Proof: We employ the Pigeonhole Principle, where the types of graphs are characterized by the degrees ofnodes, and the objects are the nodes themselves. Assume $|V| = n$, and each vertex has a degree ranging from 0 to $n - 1$.

If the graph have a node with a degree of 0 (an isolated vertex), it cannot simultaneously contain a vertex with a degree of $n - 1$. Consequently, there are at most $n - 1$ degrees which are possible. Applying the Pigeonhole Principle, we deduce that at least two nodes must have the equal degree.

Definition 3

Distance between two nodes u, v in a graph G is the length (number of edges) of shortest u to v path and is denoted by d (u, v).

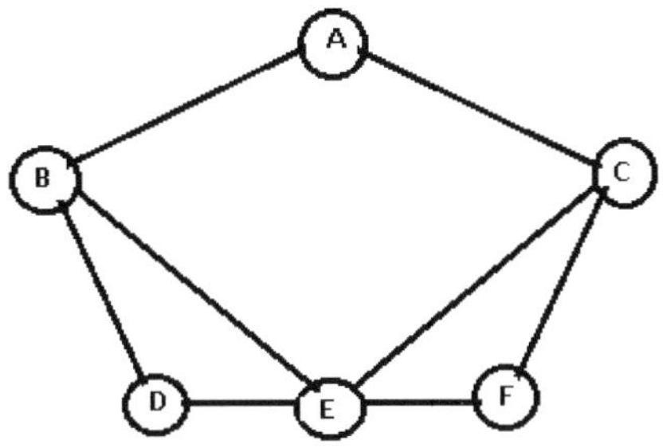

Consider the graph G in above figure.

d (A, D) = 2, d(A,E)=2, d(A,C)=1

Girth

Girth of G, denoted by $g(G)$, is the number of edges in shortest cycle of G. (If G is acyclic then its girth is denoted by w infinity).

E.g., g (G) = 3 for above graph G.

Circumference of G is the number of edges of longest cycle in graph G.

(If G is acyclic then its circumference is zero).

Diameter

Diameter of graph G is the maximum distance among all pairs of vertices

In G. Denoted by → diam (G)

diam (G) = $max_{u,\, v \in V}\, d\,(u, v)$.

Radius

Radius of graph G, denoted by rad (G), is defined as:

rad (G) =$min_{u \in V} max_{v \in V}\, d\,(u, v)$.

E.g. diam (G) =2

rad (G) = 2

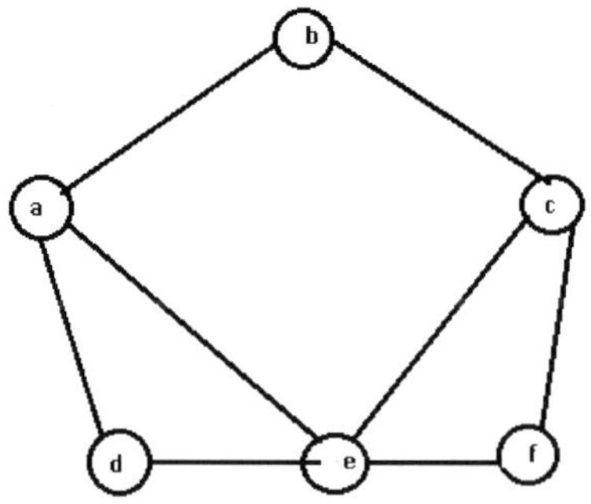

Hamiltonian Graph The graph is Hamiltonian if there is a single cycle in G that contains every vertex of G. In other words, G is Hamiltonian if and only if its circumference is |V|.

Such a cycle in G is also called a spanning cycle.

Eulerian Graph

The graph G is called Eulerian (traceable) if it has a path that connects every vertex of G.

Clique

A Clique is a subset S of the vertices V which induce a complete subgraph of G

Example 5

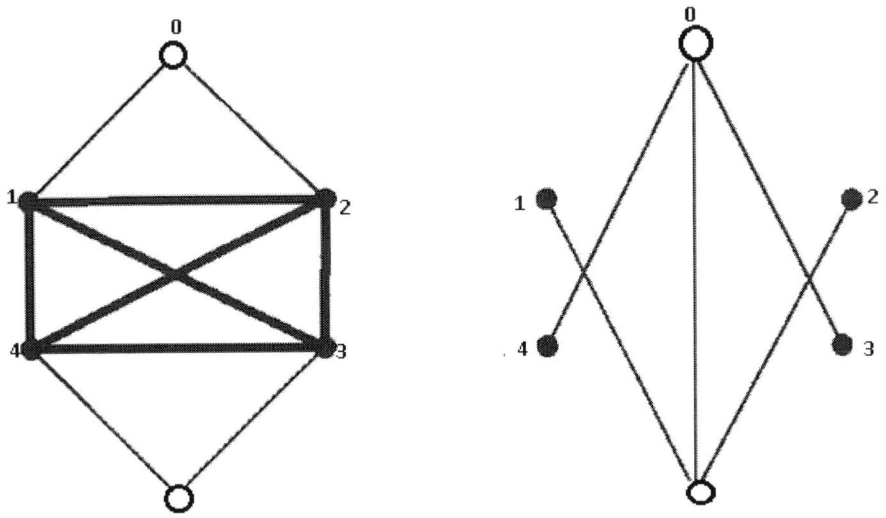

Figure: maximal clique (left) and maximal independent set (right) of a graph

Independent Set of a Graph

A subset I of vertex set of a graph G is said to be independent set if any of two vertices in I are not joined via an edge.

Consider the following subsets from the above graphs –

$S_1 = \{0, 1\}$

$S_2 = \{3,4\}$

$S_3 = \{1, 2, 3\}$

$S_4 = \{1,2,3,4\}$

Clearly, S_1 is not an independent vertex set, because both are connected vertices ,

. On the other hand, the subsets S_2, S_3, and S_4 of vertices of the graph are the independent vertex sets because the vertices in these sets are not connected.

Maximal independent set: A maximal independent set refers to an independent set in a graph that cannot be extended by incorporating additional vertices.

Put differently, this set is considered maximal because there are no external vertices that can be added without violating the independence property of the set.

Example 6

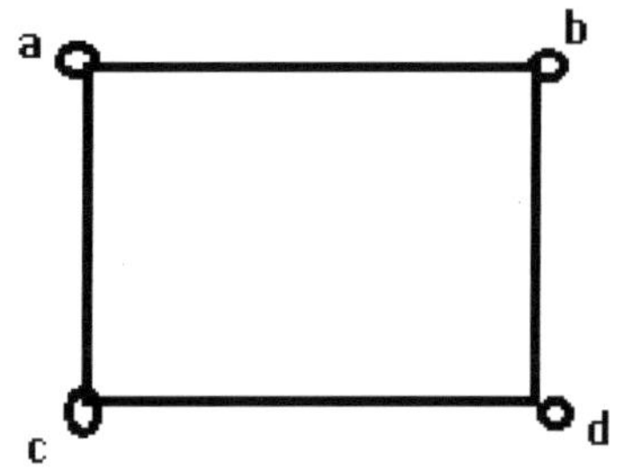

It is noted that while {A, D} and {B, C} are both maximal independent sets in above given graph. They are not the only maximal independent sets in this graph. There are other possibilities, such as {A, C}, {A, B, C}, {A, B, D}, and {B, D}. The key is that you cannot add any more vertices to these sets without violating the independence property.

Complement of a graph

The complement of a graph K is attained by constructing a graph H on the same vertices such that two distinct vertices of H are joined by an edge if and only if they don't form an edge G.

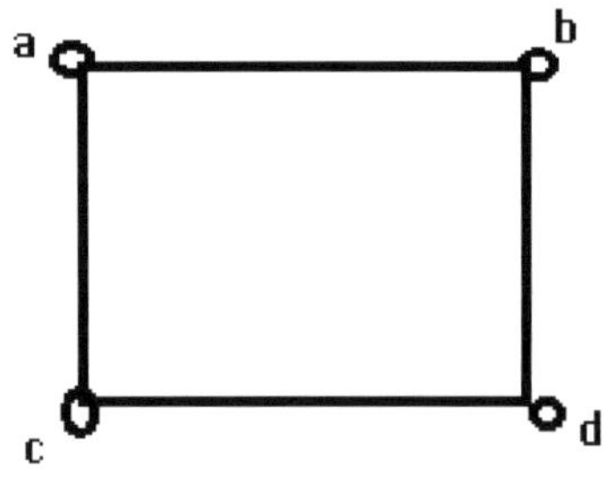

Graph G

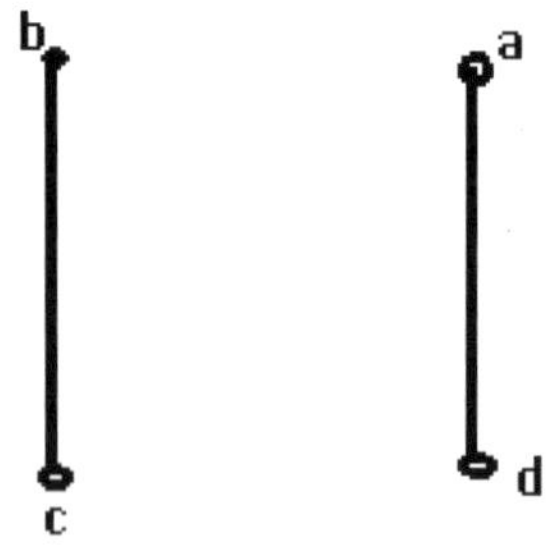

complement of graph G

Vertex cover

A subset S of the vertices V is termed as vertex cover of a graph $G = (V, E)$ so that for every $eE\in$, there exists a $v\ S\in$ such that e is attached to V. (every edge of G has at least one endpoint vertex cover).

Example 7

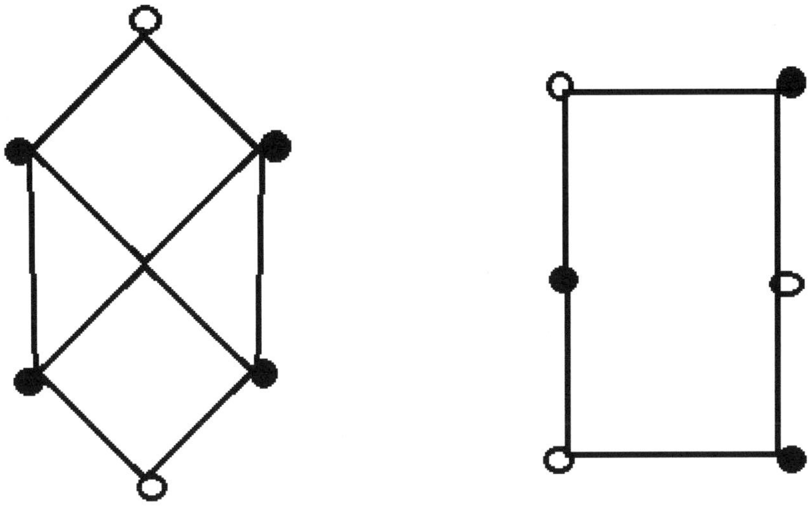

The black vertices show vertex cover of above graphs.

Theorem 2

A set I is indpendent set in a graph $G = (V, E)$ if and only if the set $V\backslash I$ is a covering of G.

Proof: Suppose I is an idempotent set. Let $\{V, V'\}\ E(G)\ \in$if $V\ I\in$, then $V'V\backslash I\ \in$ because I is idempotent set. Similarly, if $V'\ I\in$ then $v\ V\backslash I\in$ because I is idempotent. If $V, V'\ I, V, V'V\backslash I\notin\in$. so $V\backslash I$ is a covering of G.

Conversely

Let $V\backslash I$ is a vector cover of G. Let $v, v'\ I\in$. then $\{V, V'\}E(G)\notin$., then V or $V'\ V\backslash I\ \in$which gives contradiction.

This implies that I is an idempotent set.

Eccentricity of vertex

Eccentricity of a vertex u is length or number of edges of a maximal path starting from u. It is symbolised by e(u).

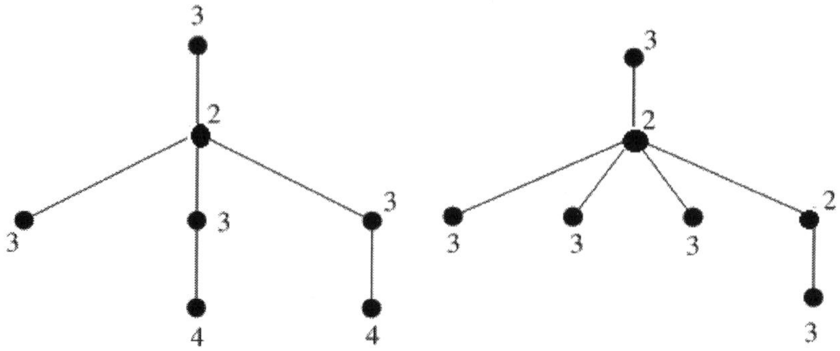

Edge induced subgraph

An edge-induced subgraph refers to a subset of edges within a graph G, inclusive of any vertices that serve as their endpoints.

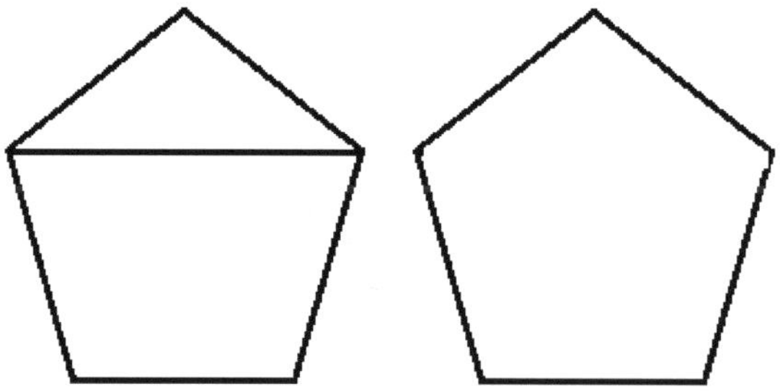

Vertex induced subgraph

For a graph G = (V, E), if V' is a subset of V, then the vertex induced subgraph H = (V', E') is defined, where $\{v_i, v_j\} \in E'$ if and only if vertices v_i and v_j are connected by an edge in G.

Example 8

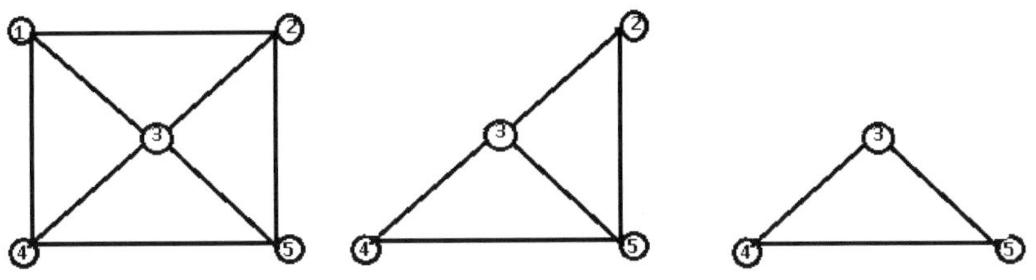

Edge deletion graph: Let $G = (V, E)$ be a graph. let $E' \subseteq E$ then graph G' attained by eliminating the edges in E' from G is a subgraph induced by $E \backslash E'$. It is called edge deletion graph.

Vertex deletion graph:

Let V⊆V′ then graph G′, obtained via eliminating vertices V′ from the graph, is subgraph induced by V\V′ and is called vertex deletion graph.

Importance of Degree Sequence in Graph theory

The degree sequence of a graph is a list that records the degree (number of edges incident to a vertex) of each vertex in the graph. Understanding the degree sequence is crucial in graph theory, as it provides valuable information about the structure and properties of the graph. Here are some key aspects that highlight the importance of the degree sequence in graph theory:

1. **Graph Classification:**
   - The degree sequence is a fundamental characteristic used to classify graphs. Graphs which have identical degree sequence are said to be "degree-preserving" or "degree-sequence graphs." The degree sequence alone can often determine certain graph properties, though not always the entire structure.
2. **Euler's Theorems:**
   - The degree sequence plays a significant role in Euler's theorems. Euler's theorem for connected graphs depicts that such graphs has exactly zero or

two vertices with an odd degree. This insight relies heavily on the analysis of the degree sequence.

3. **Handshaking Lemma:**
    - The degree sequence is linked to the Handshaking Lemma, which establishes link between the sum of the degrees of all vertices to the the number of edges in a simple graph. This lemma is a fundamental result in graph theory and provides a useful tool for analyzing relationships between degrees and edges in a graph.
4. **Graph Connectivity:**
    - The degree sequence is closely related to the connectivity of a graph. For example, if a graph has a high degree sequence, it might be more likely to be connected. The degree sequence also provides insights into the potential presence of isolated vertices or components.
5. **Graph Reconstruction:**
    - In some cases, a graph can be reconstructed uniquely (up to isomorphism) from its degree sequence. This property is utilized in graph reconstruction problems where the goal is to determine if a graph can be uniquely reconstructed from partial information.
6. **Network Analysis:**
    - In the context of social networks and communication networks, the degree sequence provides insights into the importance of nodes. Nodes with high degrees are often considered hubs or influential vertices, while nodes with low degrees may be peripheral or less influential.
7. **Algorithmic Applications:**
    - The degree sequence is used in various algorithms and processes related to graph analysis, such as designing efficient algorithms for generating random graphs with a given degree sequence or for finding certain graph properties.

Understanding and analyzing the degree sequence is, therefore, a fundamental step in unraveling the structural and connectivity characteristics of a graph, and it serves as a cornerstone for further investigations and applications in graph theory and related fields.

Degree Sequence:

Let G be a graph with IVI=n vertices. The degree sequence of G in a triple dϵZ^n such that $d = (d_1,d_2,...,d_n)$,$where\ d_i\ =\ deg(v_i)$, $\forall\ v\epsilon V(G)\ and\ d_1 \geq d_2 \geq ... \geq d_n$

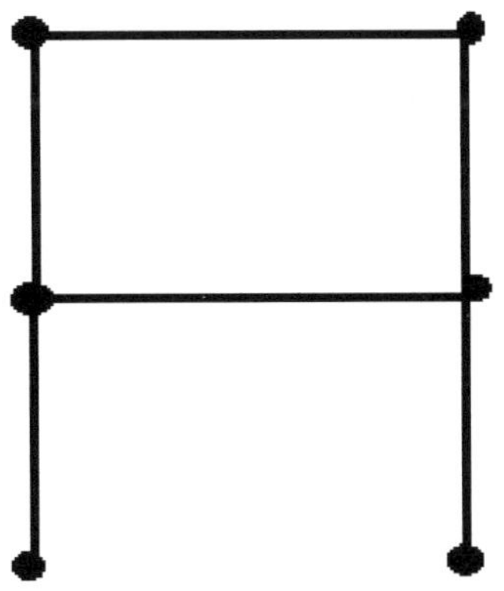

For instance, degree sequence of above graph is (3,3,2,2,1,1).

Graphic Sequence:

Let $d = (d_1,d_2,...,d_n)$, be a vector in Z^n with $d_1 \geq d_2 \geq ... \geq d_n$. Then d is said to be graphic if there exist a graph with this degree sequence $d = (d_1,d_2,...,d_n)$.

Lemma 1

For a graphic degree sequence $d = (d_1,d_2,...,d_n)$ there exist a graph G = (V, E) with a degree sequence d such that for $V = (V_1,V_2,...,V_n)$,

1. Deg (V_i) = d_i, i = 1,2, ...
2. V_1 $is\ adjacent\ to$ $V_2,V_3,...,V_{d_1+1}$.

Proof:

Given that d is graphic, so ∃ a graph having this vector d as degree sequence.

Choose G = (V, E) which have maximum value of r, where

$r\ = N(V_1) \cap \{V_2,V_3,...,V_{d_1+1}\}$.

Where, $N(V_1)$ is the neighborhood of V_1.

r shows maximum number of vertices in {$V_2,V_3,...,V_{d_1+1}\}$.

Which are neighbors of V_1.

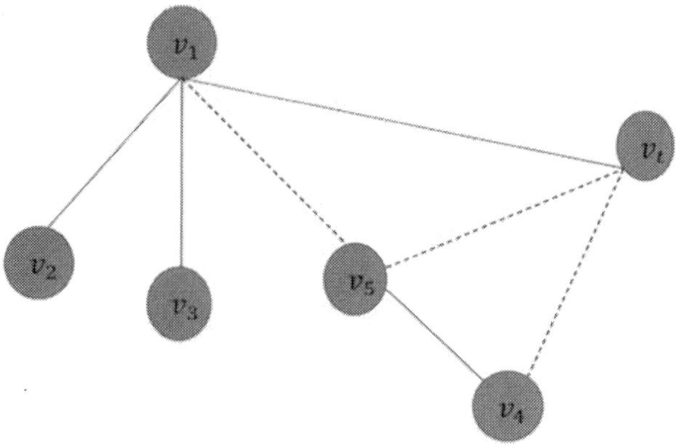

If $r = d_1$, then theorem is proved.

Let $r < d_1$

REASONS:

1. Since $deg\ (V_1)\ =\ d_1$, so there must be vertex V_t with $t\ >\ d_1$, *so that*

V is adjacent to V_t.

2. There exists a vertex V_s, $2 \leq s \leq d_1 + 1$ that has no edge incident to V_1.

By the arrangement of vertices in V, $deg\ (V_s)\ \geq\ deg\ (V_t)$

Therefore, there exist some vertex $V_k\ \epsilon V$ such that V_k is joined to V_s via an edge but not V_t.

We form a different graph $G^{'}\ =\ (V, E^{'})$.

The edges set $E^{'}$ is constructed from E by

Deleting edges $\{V_1, V_t\}$

Deleting edges $\{V_3, V_k\}$

Adding edges $\{V_1, V_s\}$

Adding edges $\{V_k, V_t\}$

Degree of V_1, V_t, V_s, V_k is preserved.

In graph G', $r' = N(V_1) \cap \{V_2, V_3, \ldots, V_{d_i+1}\}$

Here $r' > r$.

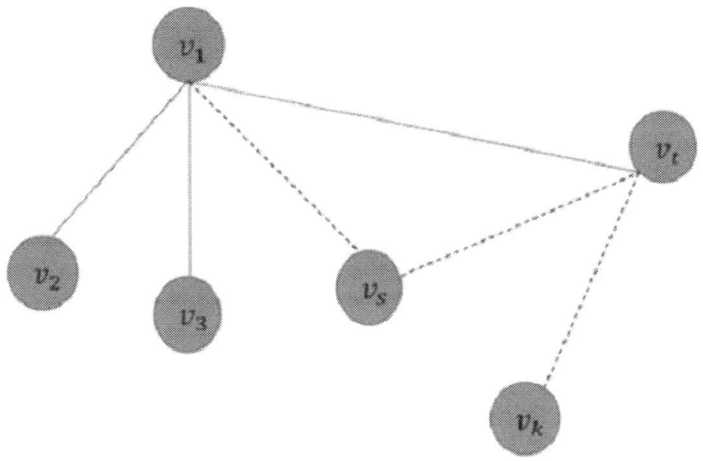

This is contradiction to our initial choose of G.

Here $r = d_1$.

Hence V_1 has an edge with $\{V_2, V_3, \ldots, V_{d_1+1}\}$.

Havel Hakimi Theorem

The Havel–Hakimi theorem provides a recursive method for constructing simple graphs from a given degree sequence. This theorem has applications in network design, coding theory, and various other areas where understanding and constructing graphs with specific degree sequences are important.

The Havel–Hakimi theorem is a result in graph theory that provides a method for constructing simple graphs based on a given degree sequence. The degree sequence of a graph is a list that records the degrees (number of edges incident to each vertex) in non-increasing order. The Havel–Hakimi theorem is particularly useful for verifying if a given sequence is graphical, i.e., whether it corresponds to the degrees of the vertices in a simple graph.

The theorem is named after mathematicians Karel Havel and Stanislav Hakimi, who independently contributed to its development. The Havel–Hakimi theorem states the following:

Theorem 3 (Havel-Hakimi)

The non-increasing sequence of non-negative integers $(d_1,d_2,...,d_n)$ is graphic if and only if

$\left(d_2-1,d_3-1,..., d_{d_1+1}-1, d_{d_1+2},d_{d_1+3} ...,d_n\right)$ is graphic.

Proof: Let $d=(d_1,d_2,...,d_n)$ is graphic. Previous lemma shows the existence of a graph G having degree sequence d, so that $d(v_i)=d_i$ for all $i=1,2,...,n$ and V_1 is linked to $V_2,V_3,...,V_{d_1+1}$ via an edge. If we delete V_1 and all incident edges of V_1 from this graph G_1 to get new graph $G^{'}$.

Now, due to chosen structure of G $V_1$is not attached to $V_{d_1+1},...,V_n$.

Then $G^{'}$has a degree sequence $(d_2-1,d_3-1,...,d_{d_1+1}-1,d_{d_1+2},...,d_n)$ and thus this degree sequence is graphic.

Conversely;

Now suppose $(d_2-1,d_3-1,...,d_{d_1+1}-1,d_{d_1+2},...,d_n)$ is graphic, giving a sure graph G with this degree sequence. One can create a new graph K that have degree sequence $d=(d_1,d_2,...,d_n)$.

Chapter 2

Planar and non-Planar Graphs

Planar graphs, a captivating subset of graphs in graph theory, hold immense importance and find applications in various fields. One of their fundamental features is that they can be drawn on a flat surface without any edges crossing, a property that has direct applications in circuit design, map representation, and network layout planning. The study of planar graphs has yielded crucial insights into graph theory, leading to the development of Kuratowski's theorem, which characterizes planar graphs by excluding certain graph minors. Planar graphs play a pivotal role in geographic information systems, where they aid in representing maps without overlapping edges, facilitating efficient route planning and geographic analysis. Moreover, in the realm of VLSI (Very Large-Scale Integration) design, planar graphs are utilized to model the layout of electronic components on a chip, ensuring that connections can be established without interference. The importance of planar graphs extends to algorithmic efficiency, as numerous problems are often easier to solve on planar graphs compared to their non-planar counterparts. Overall, the significance and applications of planar graphs underscore their utility as a foundational concept in graph theory with practical implications in diverse areas of computer science and engineering.

Definition 1

A graph is called planar if it can be represented in plane such that no edges crossing each other. Such a presentation is termed as planar representation of graph.

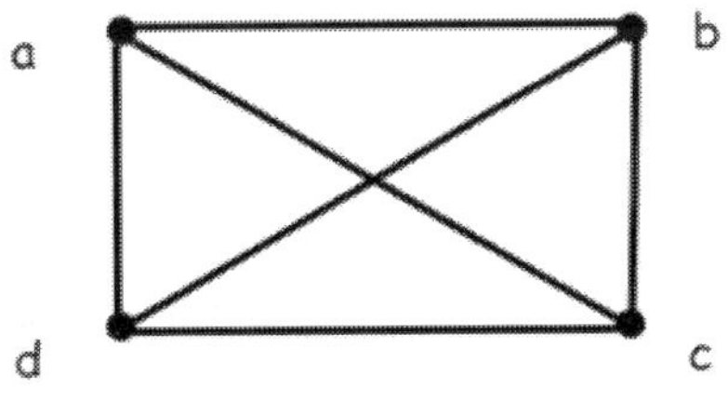

G1

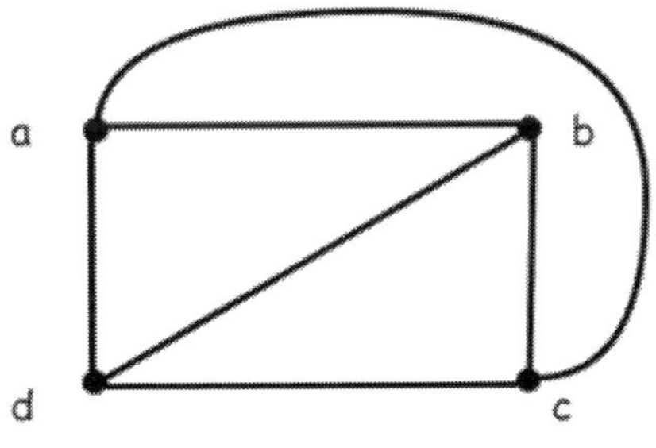

G2

So, the given graph is planar.

Theorem 1

Consider a planar connected graph G having e edges and v vertices. Then number of the region in planar representation of G, denoted by r, is

$r = e - v + 2$

Proof

We aim to prove Euler's formula for a planar connected graph via induction on number od edges , which states that in any planar representation of G, the number of regions r is given by the formula r = e - v + 2, where e denoted number of edges and v denotes number of vertices.

Base Case (e = 0): For a graph with no edges, there is only one vertex, and the only region is the entire plane. Thus, r = 1, and the formula holds: 0 - 1 + 2 = 1.

Inductive Step: Suppose the Euler formula is true for a connected planar graph with k edges, where k ≥ 0. Now, consider adding a new edge to the graph, forming a graph G' with k + 1 edges.

Adding an edge to the graph increases the number of regions by 1. We can consider this new edge as creating a new region within the graph. Therefore, the number of regions r' for G' is r + 1.

Additionally, adding a new edge introduces two new vertices connected by this edge. Hence, the number of vertices increases by 2. Therefore, the number of vertices v' for G' is v + 2.

By the induction hypothesis, the number of regions r in G is given by r = (k - v) + 2.

Now, for G', the number of regions r' is given by:

$r'=(k+1)-(v+2)+2k-v+1+2r'= (k-v)+2+1$

$r'=(k-v)+2+1$

Now, substitute the expression for r from the induction hypothesis:

$r'=r+1$

This completes the induction step.

Conclusion: By mathematical induction, Euler's formula r = e - v + 2 holds for any connected planar graph G with e edges and v vertices.

Degree of a region

If an edge appears twice on the boundary of a region, it contributes two to the degree of the region.

Degree of a region is a number that counts edges making the boundary of the region. Consider a connected graph having 6 vertices and 13 edges. A planar graph have 9 regions by Euler's formula.

Remark : Sum of degree of regions =2|E|

Theorem 2

If G is connected planar simple graph with e edges and v vertices, v > 3, then

E < 3v-6

Proof:

Each region in G has degree ≥ 3,

$$2e = \sum \deg(R) \geq 3r$$

Then by Euler Formula

$$r = e - v + 2$$

$$2e = \sum \deg(R) \geq 3(e - v + 2)$$

$$2e \geq 3e = 3v + 6$$

$e \leq 3v - 6$

Corollary 1

In the case of a connected planar graph G, there exists a vertex with a degree of at most 5.

Proof

The statement holds for a graph having at most two vertices. If G has least 3 vertices, then by previous therom. $e \leq 3v - 6$,

$$2e \leq 6v - 12 < 6v \quad (1)$$

We prove statement by contrapositive method, if degree

$$2e = \sum_{v \in V} d(V) \geq 6v$$

$$2e \geq v \qquad (2)$$

From (1) and (2), we get a contradiction. It proves there exists a vertex having degree not greater than 5.

Corollary 2

For simply connected planar graph having x edges and y vertices y ≥ 3 and havingno cycle of length 3, then x ≤ 2y – 4

Proof: Since each region in G has degree at least 4

$$2x = \sum \deg(R) \geq 4.r$$

$$2x \geq 4r$$

$$2x \geq 4(x - y + 2)$$

$$2x \geq 4e - 4y + 8$$

$$-2x \geq -4y + 8$$

$$x \leq 2y - 4$$

Subdivision

If in a simple connected graph G, we delete an edge $\{v,w\}$ and add a new vertex x together with edge $\{v, x\}$ and $\{x,w\}$, then this operation is called subdivision.

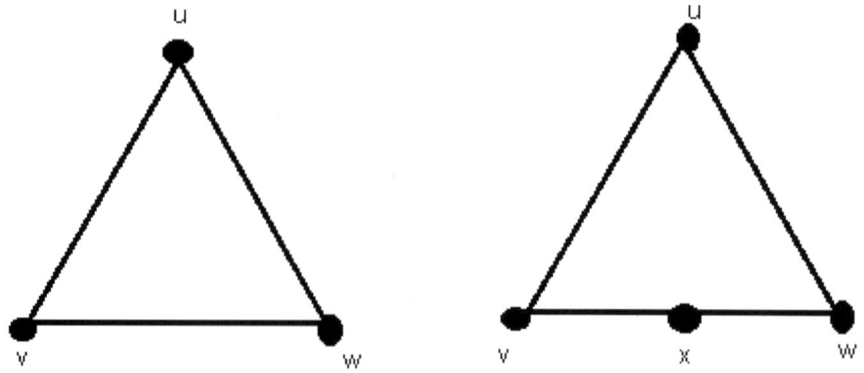

Homeomorphic graphs

A graph $G = (V, E)$ is called homeomorphic to graph $H = (V_H, E_H)$ if one can be attained from other graph after a sequence of elementary sub-divisions.

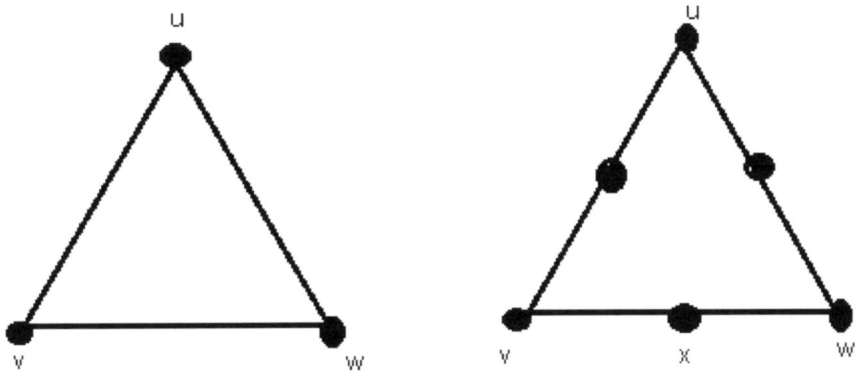

All are homeomorphic to each other.

Kuratowski's Theorem:

A graph is non-planar graph if and only if it has a subgraph homeomorphic to complete bipartite graph $K_{3,3}$ or complete graph K_5.

Proof Sketch:

Forward Direction (Planar Graphs do not contain K5 or $K_{3,3}$):

Assume G is a planar graph, and we want to show that G does not have a subgraph which is homeomorphic to K_5 or $K_{3,3}$.

1. For a planar, connected graph with v vertices, e edges, and r regions, eulers formula

$v - e + r = 2$.

2. **Edge Bound for Simple Planar Graphs:**, the number of edges e is bounded by $e \leq 3v - 6$, for planar connected graphs.
3. **Contradiction by Vertex and Edge Count:** Assume G contains a subgraph homeomorphic to K_5 or$K_{3,3}$.
   - For K_5: $v = 5, e = 10$. Plugging into Euler's formula, $5 - 10 + r = 2$ implies $r = 7$.
   - For $K_{3,3}$: $v = 6, e = 9$. Plugging into Euler's formula, $6 - 9 + r = 2$ implies $r = 5$.

   Now, using the edge bound for simple planar graphs, we find:

   - For K_5: e ≤ 3v−6e ≤ 3v−6 implies 10 ≤ 9, which is a contradiction.
   - For $K_{3,3}$: e ≤ 3v−6e ≤ 3v−6 implies 9 ≤12, which is not a contradiction.

   Since the condition for $K_{3,3}$ does not lead to a contradiction, it is possible for a graph to contain $K_{3,3}$ and still be planar. However, the condition for K_5 always leads to a contradiction.

Therefore, a planar graph don't have a homeomorphic copy of K_5 and, by extension,neither contain a copy homeomorphic to K3,3.

This completes the sketch of the proof for one direction of Kuratowski's Theorem. The other direction, proving that a non-planar graph must contain K5K5 or K3,3 involves constructing these subgraphs in a non-planar graph and is typically done by case analysis.

Example 1

Let us discuss planarity of given graph using Kuratowski's theorem.

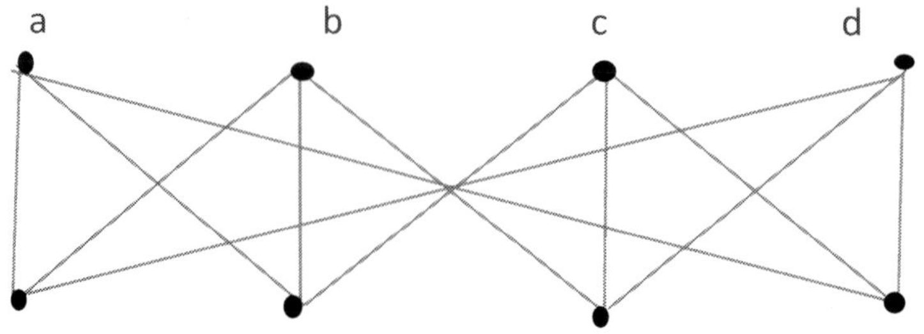

e f g h

Solution:

This graph can be drawn as

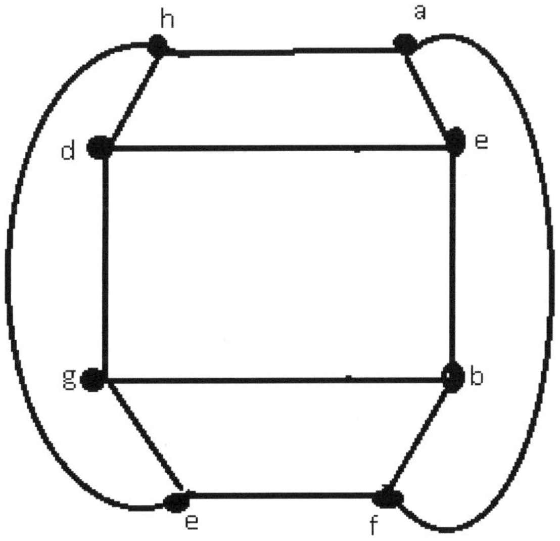

It is planar graph by Kuratowski's theorem there is no subgraph which is homeomorphic to $K_{3, 3}$

Chapter 3

Eulerian and Hamiltonian Graphs

In this chapter, we embark on a scientific inquiry into the structures and connectivity of graphs, specifically focusing on Eulerian and Hamiltonian graphs. Our investigation commences with the pioneering work of Leonhard Euler in the 18th century, who introduced the concepts of Eulerian paths and circuits. These notions, rooted in the mathematical abstraction of graphs, laid the groundwork for the formalization of graph theory.

Eulerian graphs, characterized by the existence of closed Eulerian circuits, are examined through the lens of both theoretical principles and practical applications. The exploration of Eulerian paths and circuits contributes to a profound understanding of the relationships between vertices and edges within graph structures. As we delve into this realm, we draw connections to historical challenges such as the Seven Bridges of Königsberg and extend our analysis to contemporary applications in transportation networks and circuit design.

Transitioning seamlessly, our inquiry encompasses Hamiltonian graphs, named after William Rowan Hamilton, which introduce a complementary perspective on graph traversability. Hamiltonian graphs pose fundamental questions regarding the existence of cycles that visit each vertex exactly once, opening avenues for theoretical exploration and algorithmic considerations.

Throughout this scientific journey, we delve into foundational theorems, algorithmic methodologies, and real-world applications associated with Eulerian and Hamiltonian graphs. From a scientific standpoint, these concepts serve as crucial tools for modeling and understanding complex systems, and their implications extend beyond theoretical abstraction to practical problem-solving.

This chapter invites scientists, mathematicians, and computer scientists to engage in a rigorous exploration of paths, cycles, and connectivity within the framework of graph theory. By combining historical context with contemporary applications, we aim to foster a scientific spirit of inquiry and discovery, encouraging readers to analyze and comprehend the intricate patterns that underlie the connectivity of graphs.

Euler Circuit (Euler Tour)

An *Euler circuit* in a graph G is a circuit that uses every edge in the graph exactly once.

Eulerian Trail(Euler Path)

A trail (path) that contains exactly one copy of each edge in E is called Eulerian trail.

Example 1

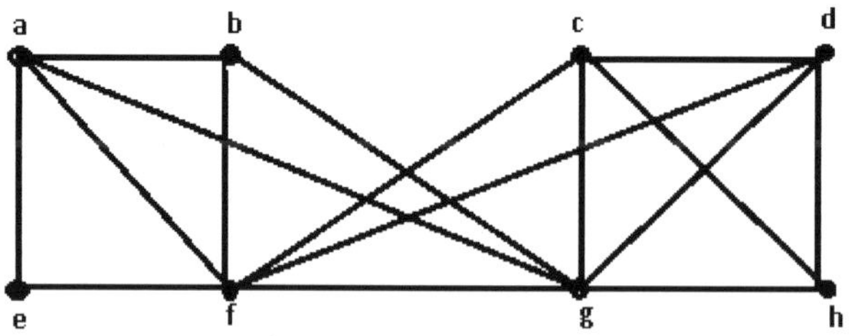

The Euler tour (a,b,g,c,d,h,c,f,d,g,h,b,f,e,a,f, g,a).

Eulerian Graph

A connected graph G is said to be Eulerian if G has a closed trail having every edge of G

Hamiltonian Path

It is a path that contains exactly one copy of each vertex in V.

Hamiltonian Circuit

It is a closed path in a graph G which contains each and every vertex of G once only.

Hamiltonian Graph

A connected graph G is called Hamiltonian graph if it has a Hamiltonian circuit.

Algorithm (Euler Circuit), Fleury's Algorithm

Theorem 1

A connected multi graph having minimum two vertices has a Euler circuit if and only if each node has even degree.

Algorithm:

Input G: Connected multi graph with all vertices of even degree.

Circuit: A circuit in G starts at an arbitrarily selected node, with edges added stepwise

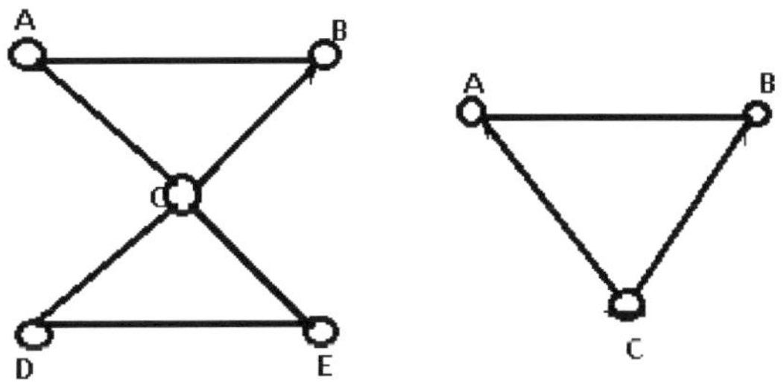

to form a path that ends at this vertex.

H: G with edges of then circuit removed while H has edges.

Begin:

Sub circuit:

A circuit in H that is also in end point of an edge of circuit.

H: = H with edges o subcircuit and all isolated vertices removed.

Circuit: = circuit together with subcircuit added at the appropriated vertex.

End {circuit is a Euler circuit}

Euler Paths and Euler Circuits:

A path that has every edge of a graph exactly once is termed as Euler path.

A circuit that contains every edge of a graph once only is termed as an Euler circuit.

Starting and ending vertices are distinct for an Euler path.

A Euler circuit starts and ends at the same vertex.

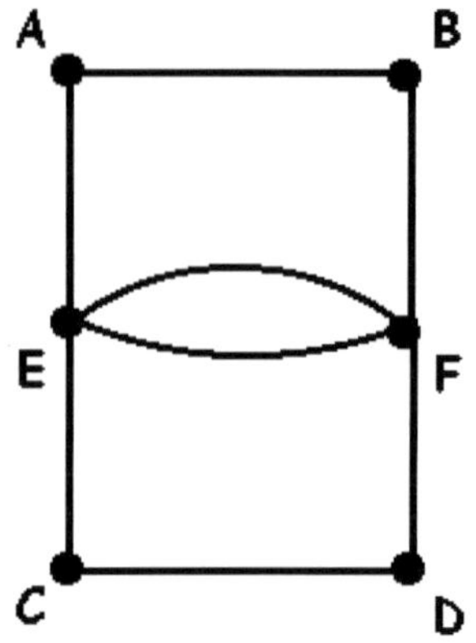

Fleury's Algorithm:

It is applied to determine a Euler circuit in a Euler graph.

Step: 1 Choose an arbitrary vertex V_0 in graph G and says walk T_0 is V_0.

Step: 2 Consider the walk, V_0 e_1 V_1 e_2 V_2 ... V_{n-1} e_n V_n to obtained such that deleting the edges e in every step to form the Graph. $G_i = G_{i-1} - G_i$ where i >= 1 and G_0 =G.

In the following manner:

If there is only one edge incident on vertex V_i then we walk on those edges. If there are several edges incident on vertex V_i then we choose an edge that doesn't form a bridge.

Bridge / Cut-edge / Cut Arc.

Are those edges if we remove one edge then graph is disconnected.

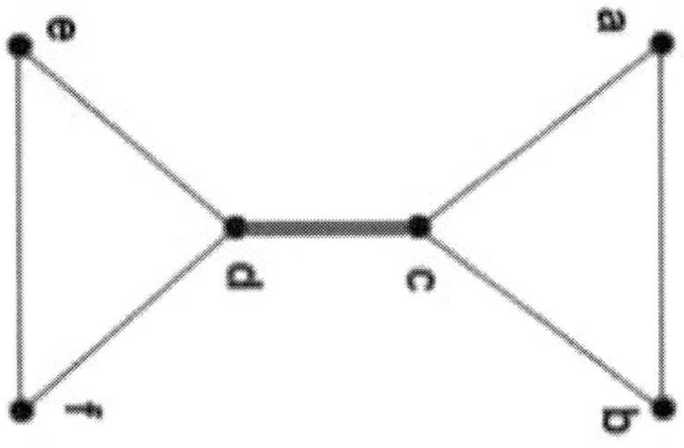

Example of Fleury's Algorithm

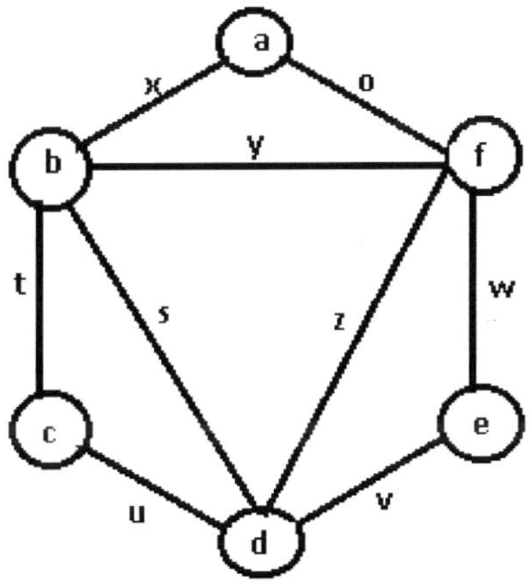

Table form of Fleury's Algorithm			
Current Walk	G_i	Next Edge	Reasoning
$T_0 = a$	$G_0 = G$	x	x is not bridge.

$T_1 = a\,x\,b$	$G_1 = G_0 - \{x\}$	y	y is not bridge.
$T_2 = a\,x\,b\;\;y\,f$	$G_2 = G_1 - \{y\}$	z	z is not bridge.
$T_3 = a\,x\,b\;\;y\,f\,z\,d$	$G_3 = G_2 - \{z\}$	s	s is not bridge but v is a bridge.
$T_4 = axbyfzdsb$	$G_4 = G_3 - \{s\}$	t	t is only edge incident on a
$T_5 = axbyfzdsbtc$	$G_5 = G_4 - \{t\}$	u	u is only edge incident on b
$T_6 = axbyfzdsbtcud$	$G_6 = G_5 - \{u\}$	v	v is only edge incident on c
$T_7 = axbyfzdsbtcudve$	$G_7 = G_6 - \{v\}$	w	w is only edge incident on d
$T_8 = axbyfzdsbtcudvewf$	$G_8 = G_7 - \{w\}$	o	o is only edge incident on e
$T_9 = axbyfzdsbtcudvewfoa$	$G_9 = G_8 - \{o\}$	……..	All edges have been traced.

Thus, by Fleury's Algorithm, T_9 is required Euler path.

ORE'S THEOREM, DIRAC'S THEOREM AND EXAMPLES

Ore's Theorem

If $deg\,(u) + deg\,(v) \geq n$, (n≥3) for every pair of nonadjacent vertices u and v in a graph G of order n, then the graph G is Hamiltonian graph.

Proof:

Consider a simple graph G with n≥3 vertices satisfying the conditions of Ore's Theorem.

Suppose that G is not Hamiltonian. Then, there exists a pair of non-adjacentnodes, say x and y, such that x and y are not part of any Hamiltonian cycle.

Since x and y do not make an edge, so deg(u)+deg(v)≥n. However, in a non-Hamiltonian graph, u and v cannot both have degree n−1 (otherwise, they would form a Hamiltonian cycle). Therefore, at least one of deg(u) or deg(v) must be less than n−1.

Assume deg(u)<n−1. Now, consider the neighbors of u. Since deg(u)<n−1, there must be at least one vertex, say w, that is not adjacent to u. Now, if we add the edge uw to the graph, we form a cycle that includes u, v, and w.

This cycle contradicts our assumption that u and v are not part of any Hamiltonian cycle. Therefore, our assumption that G is not Hamiltonian must be false. Hence, G is Hamiltonian.

It demonstrates that if a graph satisfies the degree-sum condition, then it is guaranteed to be Hamiltonian.

Dirac's Theorem

Dirac's Theorem is a classic result in graph theory, named after the Swiss mathematician Gabriel Andrew Dirac. The theorem provides a sufficient condition for a graph to be Hamiltonian, meaning it has a Hamiltonian cycle (a cycle that visits each vertex exactly once).

Theorem 2

If G is a simple graph having n nodes ($n \geq 3$) such that the degree of each node is at least $n/2$, then the graph G is Hamiltonian.

Proof:

Let G be a simple graph with n nodes and the degree of every node is greater than or equal to n/2.

On contrary assume that G is non Hamiltonian.

Ore's Condition: According to Ore's Theorem, if G is a simple graph with n nodes ($n \geq 3$) and for every pair of non-adjacent nodes u and v, the sum of their degrees is at least n, then G is Hamiltonian.

1. **Contradiction:** Assume there exist non-adjacent nodes u and v in G satisfying

$$deg(u) + deg(v) < n.$$

2. **Adding an Edge:** Add an edge between u and v. The resulting graph, denoted G', is still simple.
3. **Degree Increase:** The degrees of u and v in G' increase by 1 each.
4. **New Degrees:** The new degrees of u and v are $deg(u) + 1 \; and \; deg(v) + 1.$
5. **Comparison:** Since $deg(u) + deg(v) < n$, we have

$$(deg(u) + 1) + (deg(v) + 1) < n + 2$$

6. **Ore's Condition Holds:** However, this means that the degrees of u and v in G' satisfy Ore's Condition, contradicting our assumption that G is not Hamiltonian.
7. **Conclusion:** The assumption that G is not Hamiltonian leads to a contradiction. Therefore, if G has minimum degree at least $n/2$, it must be Hamiltonian.

The concept of Ore's Theorem and the contradiction-based approach are key elements in proving the Hamiltonicity of graphs.

Example 1

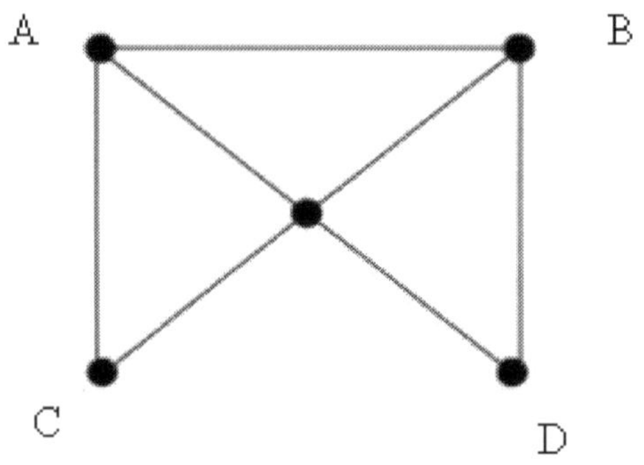

This graph does not fulfill condition of Dirac's theorem. But still this graph is Hamiltonian cycle. B D E C A B .

Example 2

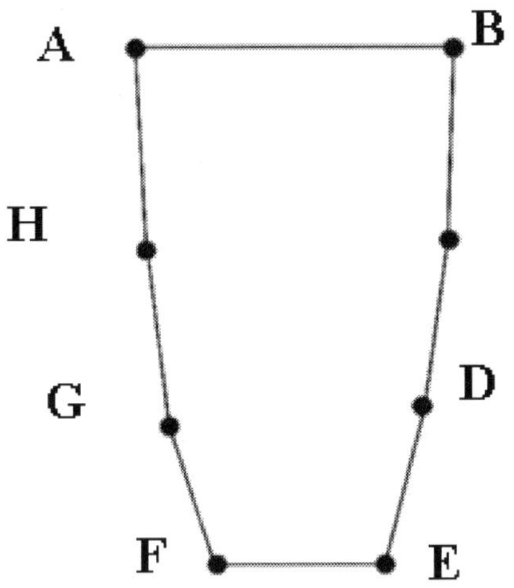

In this case every vertex has two degree and not fulfil condition of Dirac's theorem. But still this graph is Hamiltonian cycle A B C D E F G H A .

Chapter 4

Trees and Tree Traversal

In graph theory, a tree is a special kind of undirected graph that is acyclic (contains no qrfcycles). A graph is acyclic if there is no sequence of vertices such that the first and last vertices are the same, and the edges between consecutive vertices form a cycle.

Formally, a tree is defined a

s an undirected graph T that is connected and acyclic. Here are some key properties of trees:

1. **Connectedness**: Every pair of vertices in the tree is connected by exactly one path.
2. **Acyclic**: There are no cycles in the graph, meaning that it is impossible to have a path startingand ending at same node.
3. **Minimally Connected**: Deleting any edge from the tree will disconnect the graph, making it no longer connected.

A tree with n nodes will have exactly n−1 edges. This characteristic is known as being "maximally acyclic."

Trees are fundamental structures in graph theory and have various uses in computer science, including algorithms for searching, sorting, and network design. Additionally, trees play a central role in data structures such as binary trees and heaps.

Definition 1

A graph which is connected and acyclic is called as tree.

Here are some examples of trees.

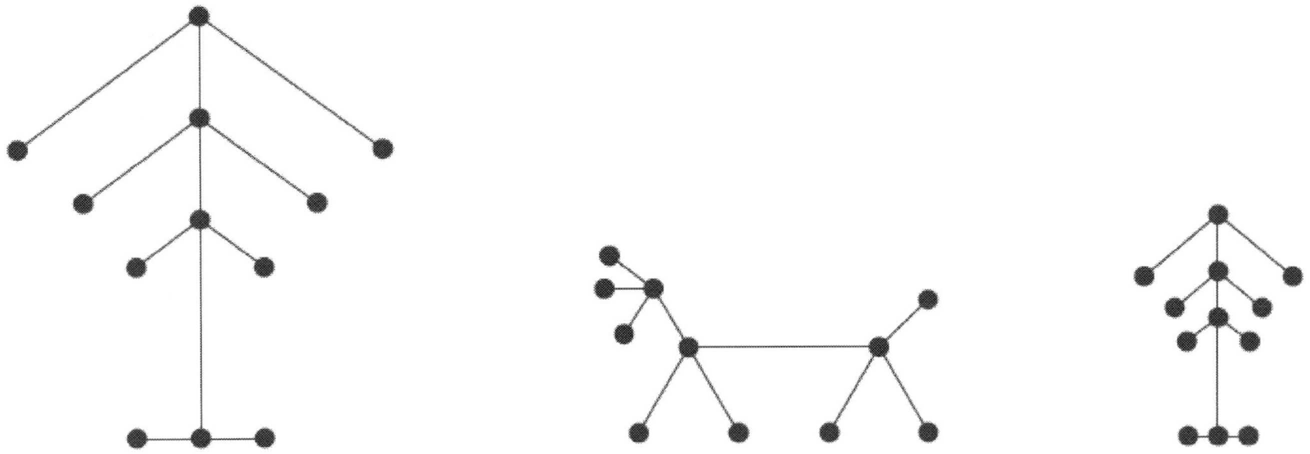

Rooted trees are commonly used in various applications, such as representing hierarchical structures, family trees, and file systems. They also play a crucial role in computer science and data structures, including in the implementation of binary trees and other tree-based algorithms.

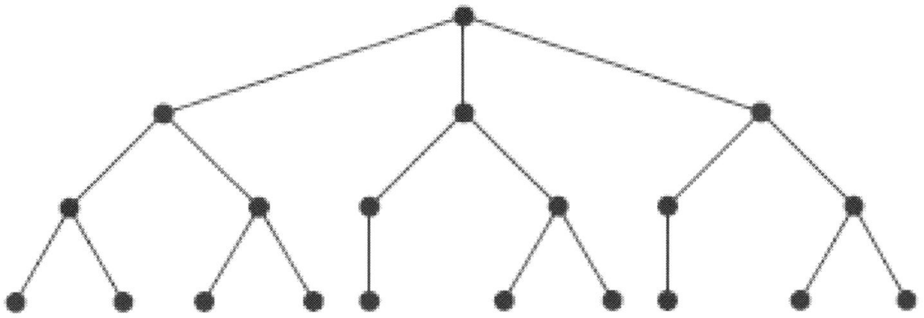

Theorem 1: If G is a Tree with vertex set V and edge set E, then $|E| = |V| - 1$.

Proof: We use induction to prove the statement.

If $|V|$=1, then $|E|$=0

$|E|$=$|V|$-1

$|E|$=1-1, so $|E|$=0.

Which is true.

Suppose theorem is true in tree for all trees with less than n vertices.

Let G be a tree with n vertices. Let $uv \epsilon E$, then $G - uv$ contain no uv -path. since uv is unique path.

Thus, $G - uv$ is disconnected so, $W(G - uv) = 2$,where W is the number of connected components.

Now, G_1 *and* G_2 , which are connected components of $G - uv$, are also trees with less than n vertices.

so, by induction $$|E(G_i)| = |V(G_i)| - 1$$

Thus, $$|E(G)| = |E(G_1)| + |E(G_2)| + 1$$

$$|E(G)| = |V(G_1)| - 1 + |V(G_2)| - 1 + 1$$

$$|E(G)| = |V(G_1)| + |V(G_2)| - 1$$

$$|E(G)| = |V(G)| - 1$$

Corollary 1

Every non trivial tree has minimally two vertices.

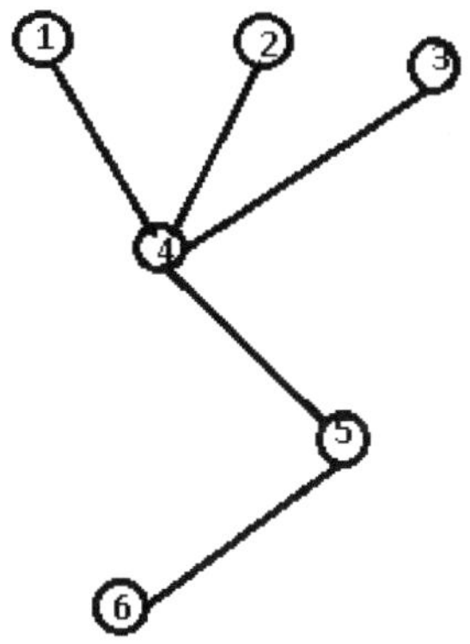

Here, in above graph, IEI=5, IVI=6

IEI=IVI-1

5=6-1

5=5.

Lemma 1

A tree having at least one edge has more than one leaf.

Proof:

A leaf in a tree is a node with a degree of 1, meaning it has only one edge incident to it.

Consider a tree with at least one edge. Let's analyze the possible cases:

1. **Case 1: Tree with two nodes (one edge):** In this case, there are two nodes, and both nodes are leaves. Thus, the statement holds true.
2. **Case 2: Tree with more than two nodes:** For a tree with more than two nodes, we know that each edge connects two nodes. Let's denote the number of nodes as nn.
    - **Base Case:** If n=2n=2, we have a tree with one edge connecting two nodes. Both nodes are leaves, and the statement holds.
    - **Inductive Step:** Assume the statement holds for a tree with kk nodes, where k>2k>2. Now, consider a tree with k+1k+1 nodes.

Since the tree is connected and acyclic, there must be at least one node with a degree of 1 (a leaf) and at least one node with a degree greater than 1 (an internal node). If all nodes had a degree of 1, it would form a chain, violating the acyclic property of a tree.

Thus, in a tree with k+1k+1 nodes, there exists at least one leaf, and the statement holds.

By induction, we have shown that a tree with at least one edge has more than one leaf.

Corollary 2

In a connected graph G if each vertex of G has degree graeter than one then G contain a cycle.

Proof:

We prove this statement on contradiction in a connected graph G if each vertex of G has degree at least two then G cannot contain a cycle. (we know that every tree with at least one edges at least two leaf) we know that in a tree a vertex has a one degree is called leaf.

Our supposition was wrong.

G contain a cycle in a connected graph with degree at least two.

Theorem 2

Consider a graph T= (V, E) with IVI=n then following statements are equivalent

i) T is tree.

ii) T is acyclic graph with n-1 edges.

iii) T is connected graph with n-1 edges.

iv) T is connected graph having each edge as cut-edge.

v) Each pair of vertices in T are joined via unique path.

vi) T is acyclic andintroducing new edge in the resulting graph precisely generates a single cycle.

Proof:

(i) Implies (ii)

If T a tree, it is by definition a connected, acyclic graph with IVI=n Now, we only need to prove that T has n-1 edges.

So, by using theorem: "If T is a tree, then IEI = IVI – 1"

No. of edges= n-1.

(ii) Implies (iii)

Suppose T is acyclic and has n-1 edges. So, we only need to prove that T is Connected. We'll use contradiction to prove this. Now Let T is disconnected then T has at least two disjoint component T_1 and T_2 (say). Each component is acyclic as T is acyclic. Now, to make a graph T connected, we need to add one edge 'e' between vertices $v_1$and v_j. When v_1 is the vertex of T_1 and v_j is the vertex of T_2. So, number of edges in $T = (n-1) + 1 = n$.

Now, G is connected and acyclic with n- vertices and n-edges. This is impossible because a connected acyclic graph is tree and tree with n-vertices has $n-1$ edges. Thus, graph T with n-vertices, $n-1$ edges and acyclic graph is connected.

(iii) Implies ((iv)

If T is connected and has n-1 edge, we need to prove that every edge in T is a cut-edge.

Prove this by contradictory method. Assume, there is an edge "e" in T that Is not a cut-edge. This means that removing edge 'e' would not disconnect the graph T. However, Since T is connected and has n-1 edges, removing any single edge will always disconnect the graph. Therefore, our assumption is false, and every edge in T must be a cut-edge.

(iv) Implies (v)

Suppose T is connected and every edge of T is cut-edge. Choose two vertices of T, *v* $\&v_0$. So, there is a path b/w v and v_o. Suppose v to v_0 are joined by at least two paths in Twhich must part ways at same point (vertex) *w* and reconnect at some other vertex w_0. One can create a new cycle in T, starting at w along first path to *w* and taking the 2nd path back to w from w_0.

Deleting any edge from this cycle cannot result in a disconnected graph. Thus, no edge in a cycle is a cut-edge contradictory to our supposition. Thus, v and v_0 are connected by a unique path.

(v) Implies (vi)

If T has cycle then there are at least two paths connecting vertices w and w' in V. This is contradiction to V. Thus, T is acyclic. Now, consider vertices v and v' and suppose that edge {v, v'} is not in T. Now there is a unique path b/w v and v', say

$$C_1 = \{v, \{v, u_1\}, u_1, \{u_1, u_2\}, u_2, \ldots\ldots\ldots\ldots\ldots\ldots, u_n, \{u_n, v'\}\}.$$

Adding edge 'e' {v, v'} to I, creates cycle

$$C_2 = \{v, \{v, u^1\}, u^1, \{u^1, u_2\}, u_2, \ldots\ldots\ldots\ldots\ldots\ldots, u_n, \{u_n, v'\}, v', \{v', v\}, v\}.$$

Hence, the creation of at least one cycle is ensured. To establish the uniqueness of this cycle, observe that in the presence of another cycle, it must include the edge {v, v'}.

Let this cycle is C_3,

$$C_3 = \{v, \{v, w_1\}, w_1, \ldots, w_n, \{w_n, v'\}, \{v', \{v', v\}, v\}$$

where, wi is present in C_3, but not in C_2, then they get two different paths joining v and v', that are

$\{v, \{v, u_1\}, u_1, \ldots, \{u_n, v\}, v'\}$

$\{v, \{v, w_1\}, w_1, \ldots, \{w_n, v'\}, v'\}$

that is contradiction to our supposition V. Thus, cycle by adding an edge is unique.

(vi) implies(i)

Suppose T is acyclic and adding an edge in T creates exactly one cycle. To show, T is tree, we will prove just that T is connected. On contrary,

Suppose T has at least two connected components.

Choose $,v' \in V(T)$, such that v and v'doesn't belong to same connected component. The edge $e = \{v, v'\}$doesn't belong to $E\ (T)$. Adding 'e' to 'T' does not create a cycle. Because 'e' is a cut-edge, so it does not lie on a cycle. This is a contradiction to (vi).

So, T is connected.

Hence, T is a tree.

Minimum Spanning Trees & Prim's Algorithm Explain with problem

Spanning tree

A graph T , a subgraph of graph G, that has all nodes connected of the graph G and is a tree called spanning tree.

Dedinition

A spanning tree in a weighted connected graph is said to be minimum spanning tree if it attains the lowest sum of edge weights. If weights of all edges in a connected graph are different, we get an unique minimum spanning tree. However, when there are multiple edges with identical weights, the graph may have multiple MSTs, all having the same minimum weight. Let's consider an example to illustrate multiple MSTs.

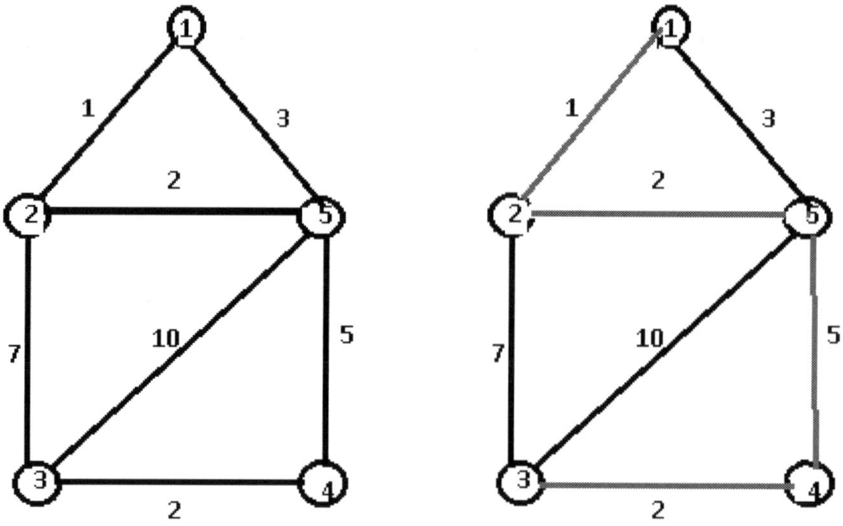

A B

In above figure, B is minimum spanning tree of graph A.

Prim's Minimum Spanning Trees Algorithm

function Prim(G: weighted connected undirected graph with n vertices)

T := a set to store the minimum spanning tree

Select an arbitrary starting vertex s

Add s to T

Repeat n - 1 times

Find the minimum weight edge (u, v) such that u is in T and v is not in T

Add v to T

Add edge (u, v) to T

end Repeat

return T as the minimum spanning tree of G

Example 1

Let's solve its example to understand it in a better way.

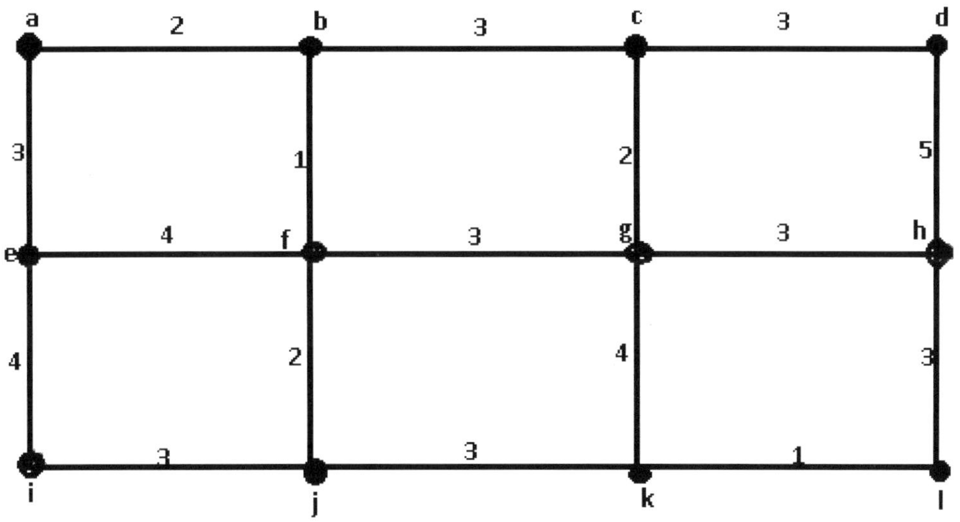

Solve by using Prim's algorithms

We choose first bf, then ab, fj, ae, ij, fg, cg, cd,gh, hl, kl and get MST with total weight sum equal to 24.

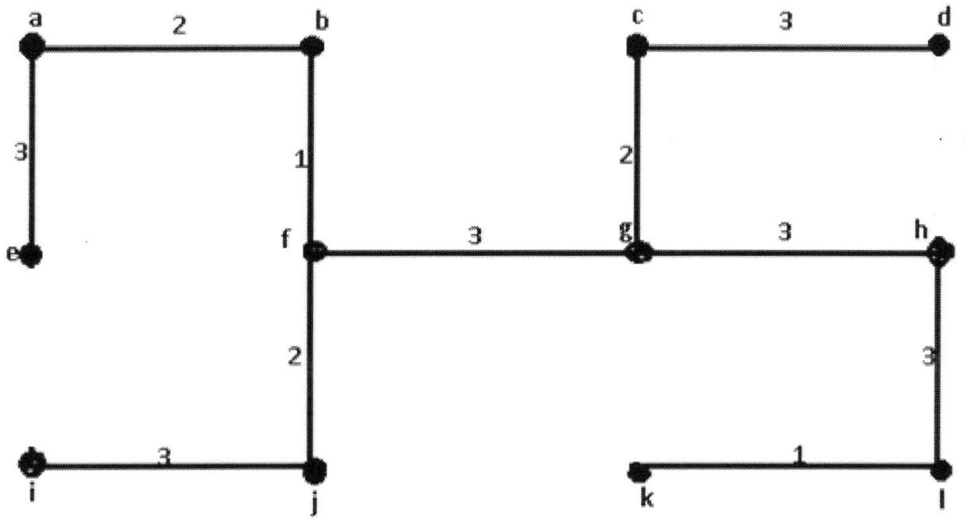

Kruskal's Algorithm:

function Kruskal (A weighted connected graph G having n nodes)

T := an initially void set

For i := 1 to n - 1

Begin

e := any edge in G with the lowest weight and don't make a cycle when added to T

T := T with e adjoined

End

Output: T is a minimum spanning tree of G)

Algorithmic Comparison

Kruskal's Algorithm:

Kruskal's greedy approach identifies a minimum spanning tree within a weighted, undirected graph. It initiates with a forest containing individual graph nodes and iteratively selects the most cost-effective edge from each node, incorporating it into the growing forest. This process persists until a single tree remains in the forest, constituting the minimum spanning tree.

This algorithm operates by taking a graph as input and determining a subset of its edges to create a tree encompassing all vertices, minimizing the total edge weight. The procedure involves sorting the graph's edges by weight, selecting the edge with the lowest weight, and adding it to the evolving tree. This sequence repeats until all vertices are encompassed in the tree.

Prim's Algorithm:

Prim's algorithm, also greedy in nature, identifies a minimum spanning tree for a weighted, undirected graph. It commences with a single node and progressively includes the most economical edge from that node into the growing tree. Unlike Kruskal's algorithm, Prim's method repeats until the tree contains n-1 edges, where n denotes the number of nodes in the graph.

Kruskal's Algorithm Complexity:

Kruskal's algorithm is a well-established technique for discovering the minimum spanning tree of a graph. As a greedy algorithm, it leverages the principle that the edges of a minimum spanning tree constitute a subset of any other spanning tree's edges.

This algorithm time complexity is O(|E|log|E|), where |E| denotes the number of edges in the graph. This complexity arises from the algorithm's use of a priority queue with a time complexity of O(log|E|). However, the algorithm's space complexity is O(E), which is relatively high.

Depth First Search(DFS) and Breadth First Search (BFS) Algorithm for Graph Traversal

DFS and BFS represent distinct approaches to graph traversal. DFS, or Depth-First Search, involves delving deeply into the graph, while BFS, or Breadth-First Search, opts for a broader exploration.

The primary contrast between these two algorithms lies in their exploration strategies. DFS systematically explores nodes directly linked to the source node before venturing into nodes indirectly connected. On the other hand, BFS initially explores nodes located at a distance of one edge from the source node, progressing to nodes at increasing distances.

Another noteworthy disparity is the order in which nodes are explored. DFS follows a depth-first order, meaning it thoroughly investigates all nodes at a particular depth before proceeding to the next depth. Conversely, BFS adheres to a breadth-first order, meaning it comprehensively examines all nodes at a specific distance from the source node before advancing to nodes at subsequent distances.

Graph Traversal

Graph traversal involves visiting and exploring every vertex in a graph as part of a processing procedure. This method is utilized for searching specific vertices within the graph and determining the sequential order in which vertices are to be visited during the search process.

Depth First Search (DFS)

1. Initialize a stack with a size equal to the total number of nodes in the graph.
2. Select any node as the initial point for traversal, visit that node, and take it over the stack.
3. Visit any adjacent node of the node at the top of the stack that has not been visited, and push it onto the stack.
4. excercise step 3 until all nodes in the stack are visited.
5. If there are no new nodes to visit, backtrack by popping one node from the stack.
6. Repeat steps 3, 4, and 5 until no node in the stack.
7. create the desired spanning tree by deleting unvisited edges from the given graph.

Example 3

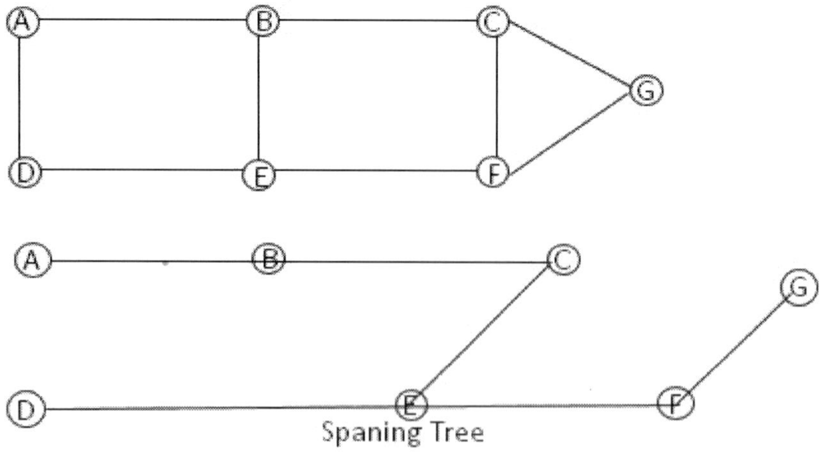

Spaning Tree

G
F
D
E
C
B
A

Output: ABCEDFG

2. **Breadth First Search**

Algorithm

Step 1. Select any vertex $v \in V$ where V is a collection of vertices and v is arbitrary vertex.

We can choose any vertex in your graph say v_1 from which you want to traverse the graph.

Step 2. Q is Queue data structure enqueue v_1 into Q. So Q is not empty.

Step 3. Add all those vertices into the data structure Q which are adjacent to vertices V.

Step 4. Remove first vertex from data structure Q. Process will continue for each and every adjacent and this process is continued until Q is empty.

Example 4

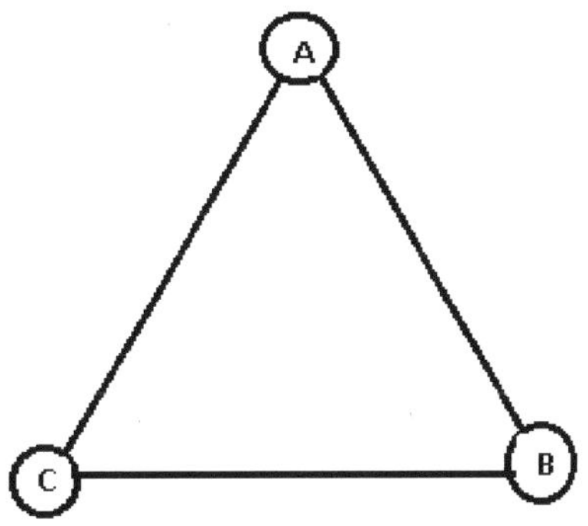

Spanning trees of this graph are as follows

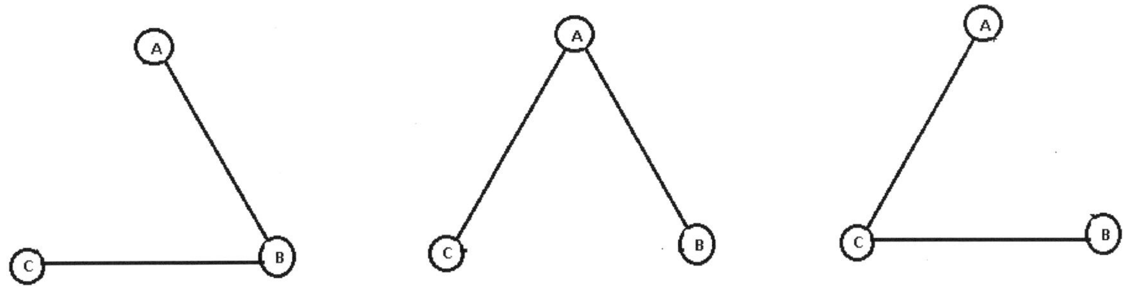

Clearly a spanning tree will have |V|-1 edges.

Example 2

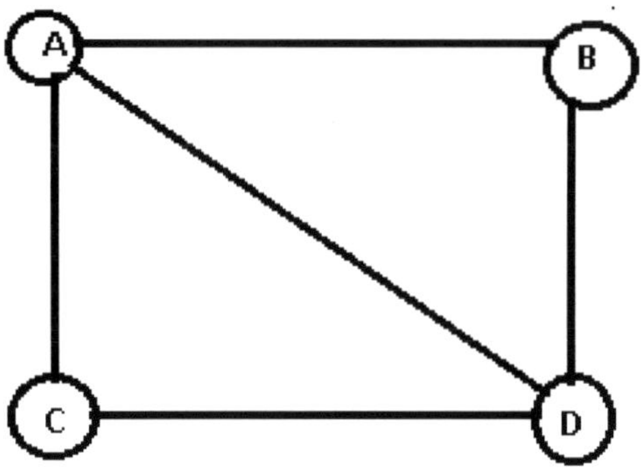

Find all spanning trees of above graph?

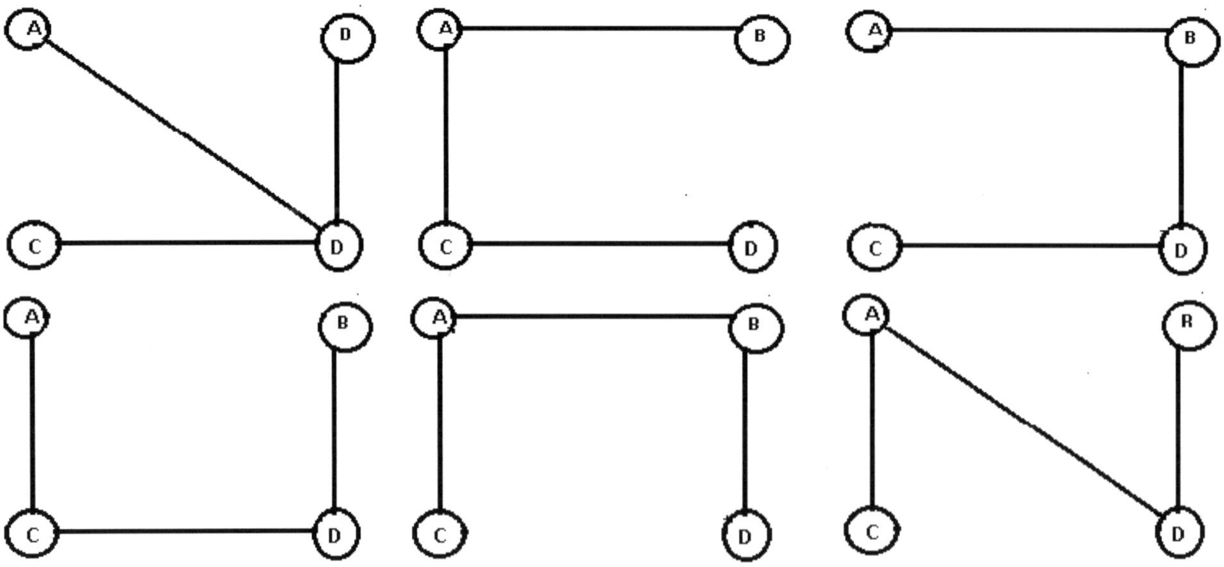

Example 5

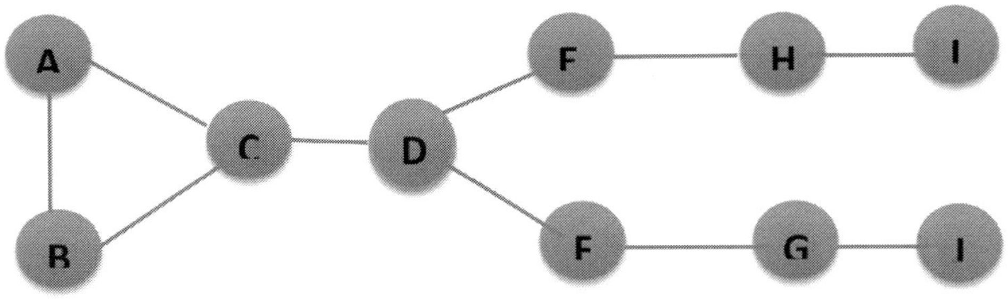

Find spanning tree of this graph by using BFs ?

Solution: We choose arbitrary vertex say, A

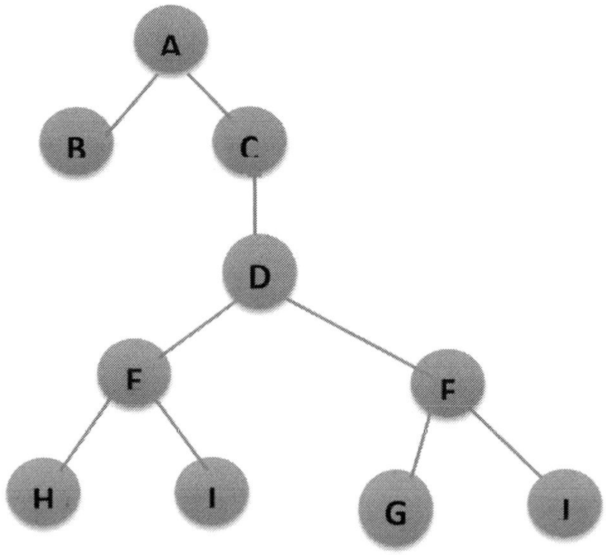

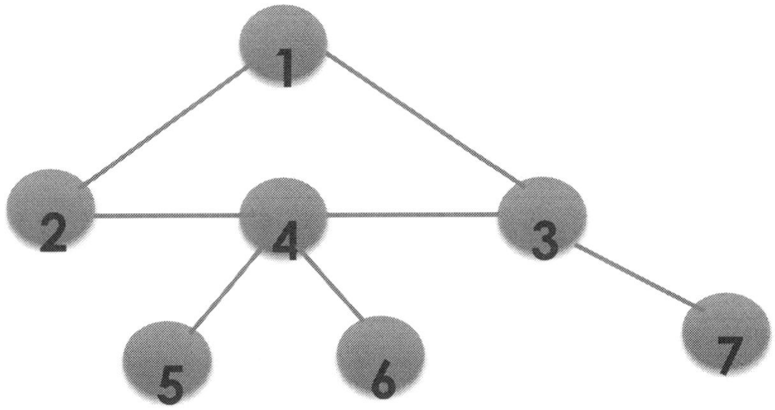

By using BFs Algorithm solve above example ?

Solution: We need that by using BFs algorithm visiting the vertices and exploring the vertices,

BFs, Q is data structure Queue

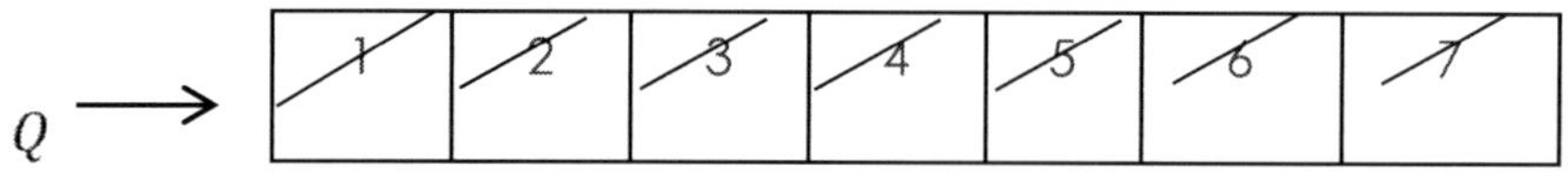

Level wise

1 2 3 4 5 6 7

Single Source Shortest Path (Dijkstra's Algorithm):

Within the expansive field of graph theory, algorithms stand as indispensable tools, unraveling the complexities inherent in intricate networks and interconnected systems. Among these algorithmic virtuosos, Dijkstra's algorithm emerges as a seminal contribution, a testament to the intellectual prowess of Dutch computer scientist Edsger W. Dijkstra. This algorithm holds particular prominence for its proficiency in addressing a foundational challenge in graph theory: the determination of the shortest path between two vertices within a weighted graph.

Here are the key steps of Dijkstra's algorithm:

1. **Initialization:**

- Set the initial vertex as the source.
- Assign zero to the source vertex and infinity for all other vertices.
- Set the initial priority queue or min-heap with all the nodes and their tentative distances.

2. **Iterative Process:**
    - Continuously iterate as long as the priority queue is not empty, and during each iteration, retrieve the vertex with the smallest tentative distance from the priority queue.
    - For each adjacent node of the current node, compute the sum of the tentative distance to the current node and the weight of the edge to that neighborind node.
    - If the sum is less than the tentative distance already given to adjacent node, change the tentative distance.
3. **Termination:**
    - Once the destination node is extracted from the priority queue, algorithm terminates, and the shortest path has been found.

Dijkstra's algorithm is employed for getting the shortest path in a weighted graph, provided that there are no negative cycles. It is often used in routing and network protocols to find the optimal path for data transmission in computer networks.

The time complexity of Dijkstra's algorithm is commonly $O((|V| + |E|) \log V)$, V and E carry usual meaning in graphs and depends on the implementation.

Example 1:

Dijkstra algorithm

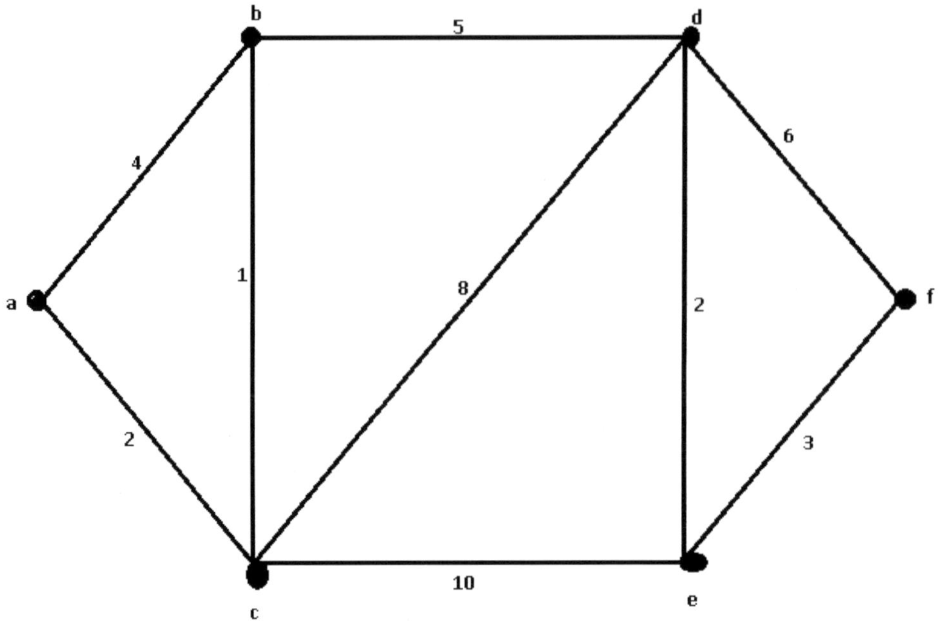

The source vertex is "a"

	a	b	c	d	e	f
a	0	∞	∞	∞	∞	∞
(a, c)		4	2	∞	∞	∞
(a, c, b)		3		10	12	∞
(a, c, b, d)				8	12	∞
(a, c, b, d, e)					10	14
(a, c, b, d, e, f)						13

Outputs:

0 2 3 8 10 13

From graph we can see that the shortest path between the vertices is given below

For node a to b, we can follow the path a-c-b. This has a distance 2+1=3, whereas the path a-b has a distance of 4. So,

a- c- b is the path with shortest distance.

Chapter 5

Graph Operations

Cartesian Product of Graphs

Let C and D represent two distinct graphs. The Cartesian product of C and D, denoted by C×D, is a simple graph with vertices in the set V(C)×V(D). For any two vertices (u_C,v_C) and (u_D ,v_D), they are adjacent if and only if:

$u_C=u_D$ and $(v_C,v_D)\in E(D)$

$v_C=v_D$ and $(u_C,u_D)\in E(C)$

If $|V(C)| = n_1$ $|V(D)| = n_2$

$|E(C)| = m_1$ $|E(D)| = m_2$

Number of vertices of C×D = n_1n_2

Number of Edges of C×D = $n_1m_2 + m_1n_2$

Example 1

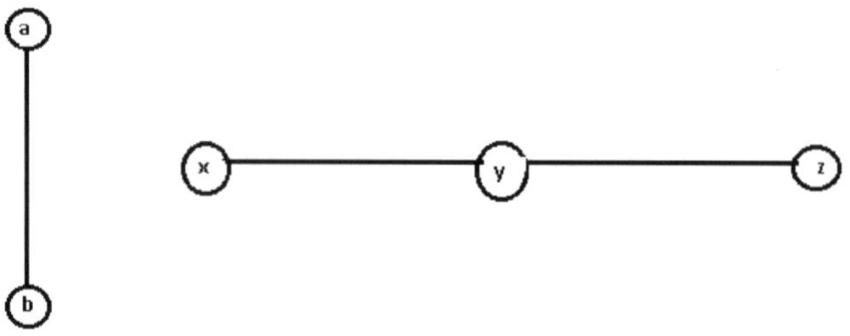

G1 G2

Solution:

$V(G_1) = \{a, b\}$, $V(G_2) = \{x, y, z\}$

$V(G_1) \times V(G_2) = \{(a, x), (b, x), (a, y), (b, y), (a, z), (b, z)\}$

And $|V(G_1)| = n_1 = 2$

$|V(G_2)| = n_2 = 3$

$|E(G_1)| = m_1 = 1$ $|E(G_2)| = m_2 = 2$

Number of vertices= $2 \times 3 = 6$

Number of edges = $4 + 3 = 7$

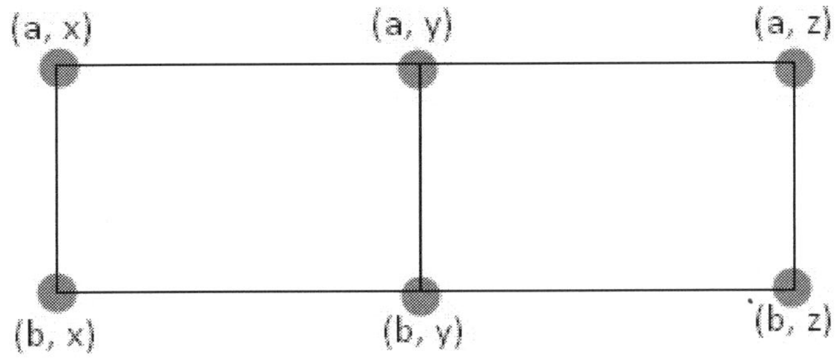

Composition of graphs:

(Lexicographic Product)

The lexicographic product of graph $G_1$and G_2 is the graph $G_1[G_2]$ having node set is V (G_1) $\times$ V (G_2) and for which (u_1 ,v_1) and (u_2 ,v_2) is an edge of $G_1[G_2]$ precisely. If

$(u_1, u_2) \epsilon E[G_1]$ or

$u_1 = u_2$ and $(v_1, v_2) \epsilon E[G_2]$

Example 2

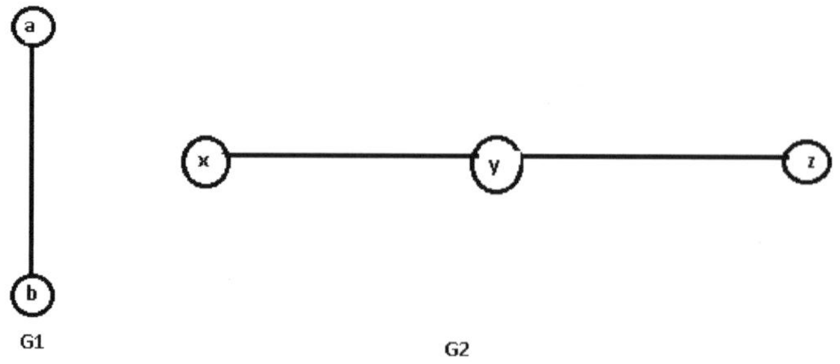

V (G_1) = {a, b}

V (G_2) = {x, y, z}

V (G_1) × V (G_2) = {(a, x), (b, x), (a, y), (a, z), (b, y), (b, z)}

$G_1 o G_2$

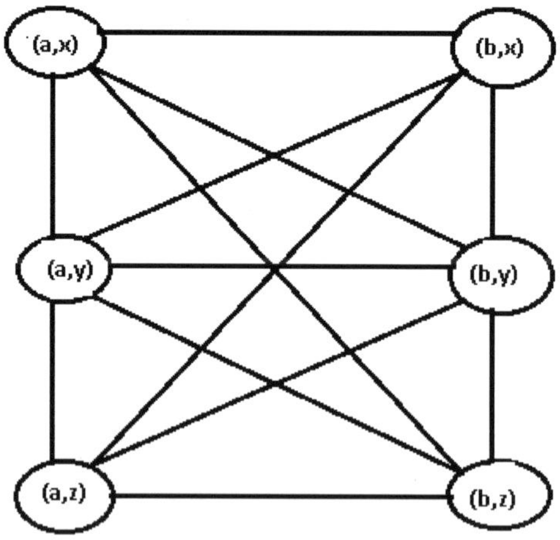

V (G_2) = {x, y, z}, V (G_1) = {a, b}

V (G_2) × V (G_1) = {(x, a), (x, b), (y, a), (y, b), (z, a), (z, b)}

$G_2 o G_1$

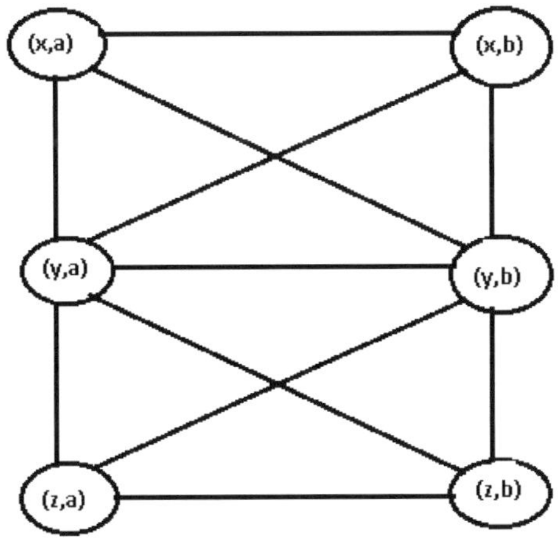

This implies that $G_1 o G_2 \neq G_2 o G_1$ composition of graph is not commutative.

The Tensor Product of Two Graphs

The tensor product of two graphs, denoted as G×H, is a binary operation defined on graphs. Given two graphs $G=(V_G,E_G)$ and $H=(V_H,E_H)$, the tensor product G×H is a graph having node set $V_{G\times H}=V_G\times V_H$, where × denotes the Cartesian product of sets.

The edge set $E_{G\times H}$ is defined as follows:

$$E_{G\times H}=\{((u_1,v_1),(u_2,v_2))|(u_1,u_2)\in E_G \text{ and } (v_1,v_2)\in E_H\}$$

In other words, for every pair of edges (u_1,u_2) in G and (v_1,v_2) in H, there is an edge between (u_1,v_1) and (u_2,v_2) in the tensor product G×H.

The tensor product is a way to combine the structural information of two graphs into a new graph. It has applications in various areas, including algebraic graph theory and network science.

Example:

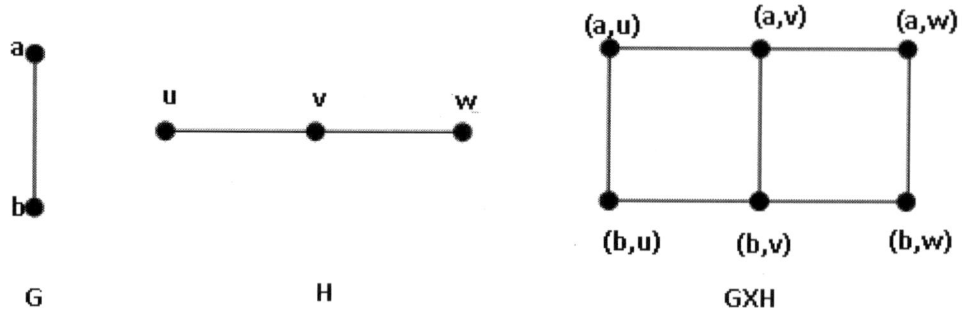

The Union of Two Graphs

The union of two graphs G and H, denoted as GUH, is a binary operation that combines the vertices and edges of both graphs into a new graph.

Let $G=(V_G,E_G)$ and $H=(V_H,E_H)$ be two graphs. The union GUH is defined as follows:

1. The vertex set of the union graph is the union of the vertex sets of GG and HH: $V_{GUH}=V_G U V_H$.
2. The edge set of the union graph is the union of the edge sets of G and H: $E_{GUH}=E_G U E_H$
3. In other words, the union of two graphs includes all the vertices and edges from both graphs. If there are any common vertices or edges between G and H, they are included only once in the union.

The union operation is a way to merge the structures of two graphs, and it forms a new graph that encompasses the combined set of vertices and edges from the original graphs.

Example 3

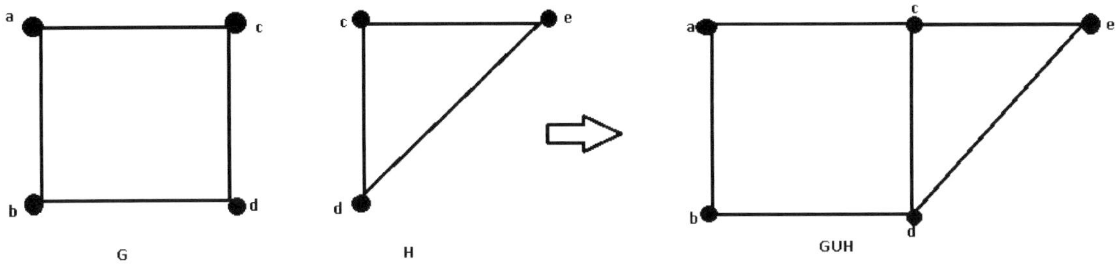

Dual Graph

In graph theory, the concept of a "dual graph" is associated with planar graphs and their geometric representations. The dual graph is derived from a planar graph in such a way that it captures certain relationships between the faces of the original graph.

Given a planar graph $G=(V,E)$ embedded in the plane, the dual graph G^* is constructed as follows:

1. **Vertex Correspondence:** Associate a vertex of the dual graph G^* with each face of the original graph G. If G has an unbounded face (exterior face), consider it as a single vertex.
2. **Edge Correspondence:** For every edge in G, introduce an edge in G^* that connects the dual vertices corresponding to the faces on either side of the original edge.
3. **Face Correspondence:** The dual graph G^* has an edge connecting two dual vertices if and only if their corresponding faces in the original graph share an edge.

The dual graph G^* provides a way to represent the adjacency relationships between faces in the original planar graph. The concept is particularly useful in the context of planar graphs and has applications in areas such as graph coloring and network design.

It's important to note that the definition of the dual graph may vary slightly depending on the specific context or the type of planar graph being considered.

Chapter 6

Graph Labeling

Within the realm of graph theory, the practice of graph labeling stands as a critical facet, where the meticulous assignment of labels to vertices, edges, or entire graphs serves as a methodological approach for encoding structural information. This inquiry seeks to delve into the profound significance of graph labeling, examining its theoretical underpinnings and diverse applications across computer science, mathematics, and interdisciplinary sciences.

Graph labeling, encompassing various schemes such as vertex labeling, edge labeling, and algorithmic approaches, holds theoretical importance in elucidating complex relationships within networks. This exploration aims to articulate the intrinsic principles of graph labeling methodologies and unveil its applications in information retrieval, network analysis, bioinformatics, and communication networks.

The theoretical foundation of graph labeling is underpinned by precise encoding methodologies, facilitating efficient information retrieval and indexing in databases. Furthermore, its application in network analysis contributes to the understanding of intricate structures, while in bioinformatics, graph labeling becomes instrumental in pattern recognition within biological systems. The logical integration of graph labeling in communication networks extends to its role in routing protocols and resource optimization.

This chapter endeavors to provide a systematic comprehension of the theoretical constructs governing graph labeling, presenting a logical framework that dissects its principles and methodologies. Whether one's background is rooted in computer science, mathematics, biology, or a related discipline, the exploration encourages an analytical engagement with the intricate language embedded in graphs and an appreciation for the expansive landscape of applications within graph labeling. Prepare for a scientifically grounded journey as we traverse the diverse dimensions of graph labeling, where every label becomes a means of deciphering patterns and unraveling complex connections.

Definition 1

Graph labeling is assignment of non-negative integers to vertices or edges of a graph,also called vertex labeling or edge labeling.

If both vertices and edges of graph are labelled under a certain rule then such a labeling is called total labeling.

Graceful Labeling

A graceful labeling of a graph, characterized by m edges, involves assigning labels to its vertices using a subset of integers ranging from 0 to m inclusively. This labeling ensures that no two vertices share the same label, and each edge is distinctly identified by the absolute difference between its endpoints. The assigned labels correspond to the magnitude of this difference, lying within the inclusive range of 1 to m. A graph that accommodates such a labeling is termed a graceful graph.

In 1967,this concept was named as "β labeling" was first introduced by a person named Alexander Rosa. Later on, it was changed to "graceful labeling" by Golomb.

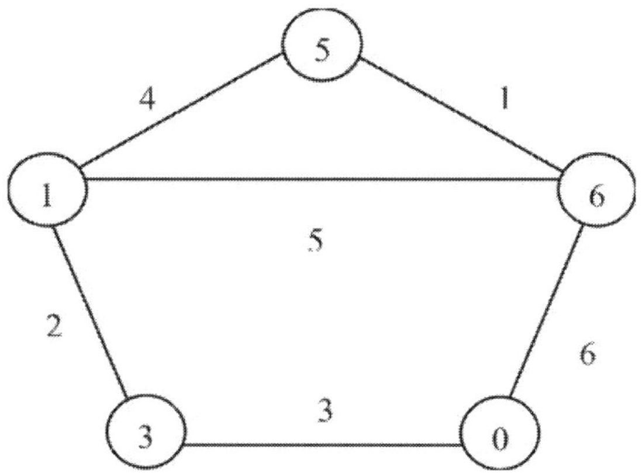

Figure : Graceful labeling of a graph.

Alpha α Labeling

The graceful labeling φ is called α labeling if an integer p exists, allowing for each edge "s_1s_2" to have in two ways $\varphi(s_1) \le p \le \varphi(s_2)$ or $\varphi(s_2) \le p \le \varphi(s_1)$. With label 1 as the edge's source, such p has to be the smaller of the other two vertex labels.

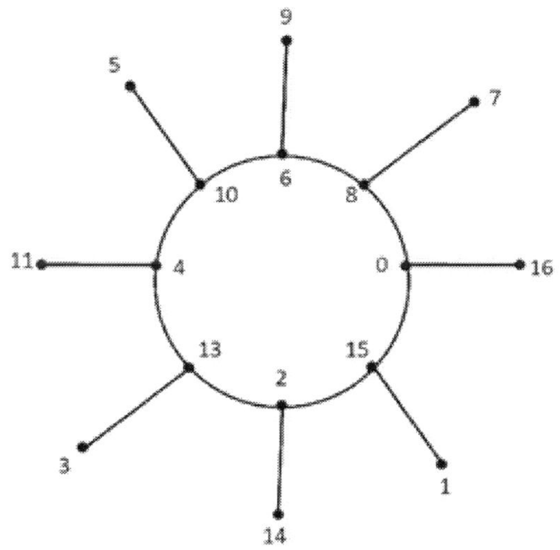

Prime Labeling

When the vertex labels of a graph can be obtained from different natural numbers {1,2,...,|V

$l\}$, and each label assigned to s_1 and s_2 is relatively prime for every edge s_1s_2, the vertex labeling is referred to as prime labeling.

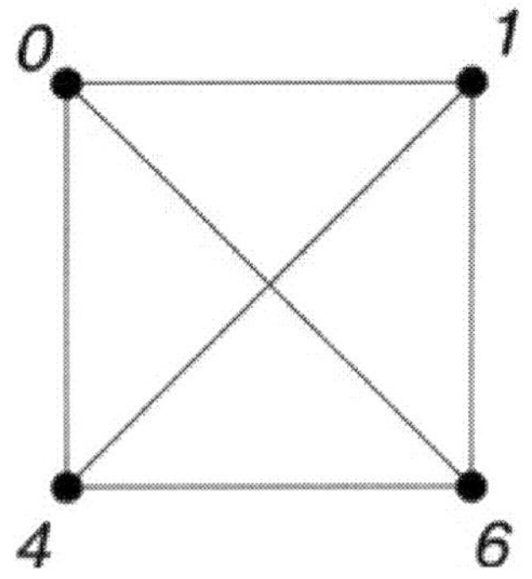

Elegant Labeling

If $G(V, E)$ is a graph having p edges, then the set $\{0, 1, 2, ..., p\}$ is the injective mapping which start which start from the vertices of G to the corresponding collection of $f(s_1) + f(s_2)$ (mode($p+1$)) and the edge label is different and non-zero.

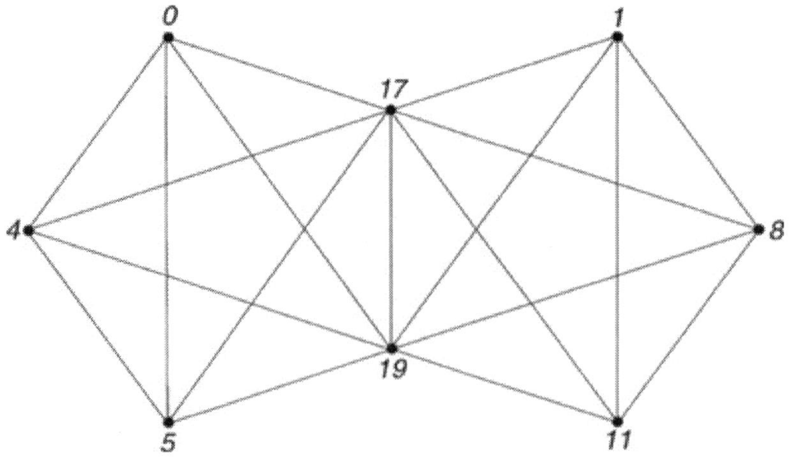

K-Graceful Labeling

If there is labeling from the graph's vertices to *{0, 1, 2, ..., s + k − 1}*, and the collection of the edge labels that results from the absolute magnitude of the difference between the labels of labouring vertices is *{k, k + 1, ..., s + k − 1}*, then the graph is considered k-graceful.

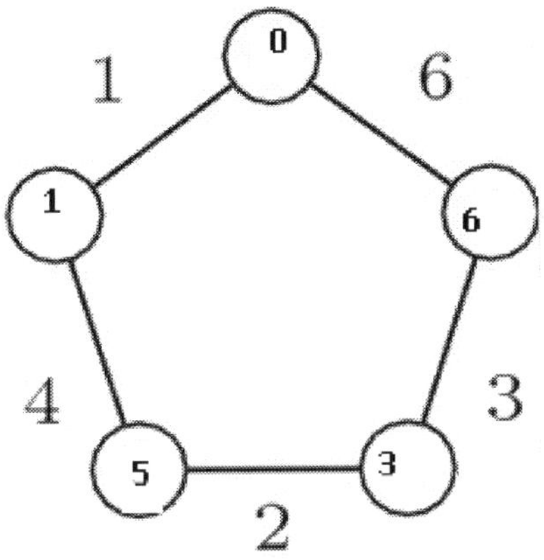

Cordial Labeling

Let ϕ denote a function mapping the vertices of the graph to the set {0, 1}. For each edge connecting vertices s1 and s2, assign the label |ϕ(s1) − ϕ(s2)|. A labeling ϕ is deemed cordial for graph G if the difference in the counts of vertices labeled 0 and 1 is at most 1, and the labels on the edges differ by no more than one.

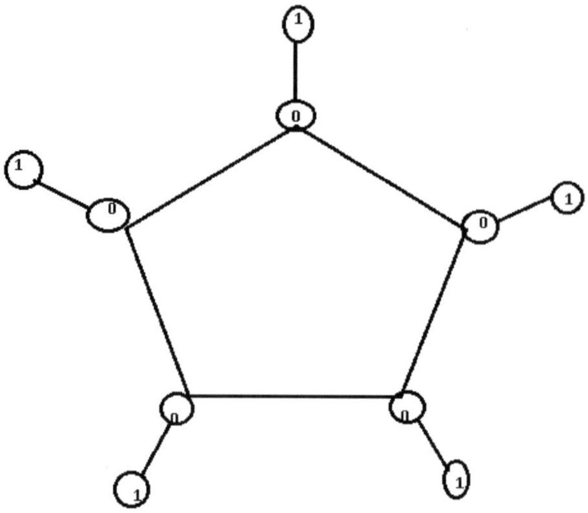

Lucky Labeling

Let $l : V(G) \rightarrow N$ represent a vertex labeling where each vertex in a graph G is assigned a positive integer. For each pair of neighboring vertices v_1 and v_2, define labeling l to be lucky labeling if $S(v_1) \neq S(v_2)$. $S(v_1)$ can be expressed as $S(v_1) = \sum_{v_i \in N(v_1)} l(v_i)$. $\eta(G)$ represents the lucky number of G, which is the smallest natural number p from the collec

tion $\{1,2,\ldots,k\}$.

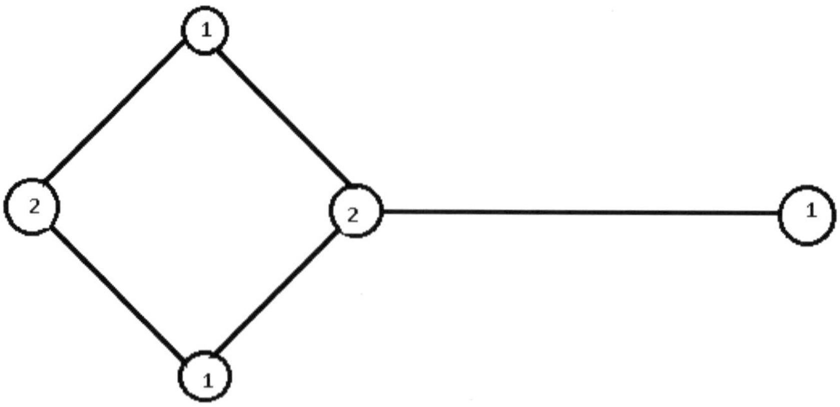

Lucky Edge Labeling

When assigning labels to the nodes of a graph using natural numbers, the edge labels E(e) are derived from the sum of the vertex labels connected by the respective edge. In this labeling scheme, known as lucky edge labeling, for any pair of adjacent edges, E(e1) is not equal to E(e2). The lucky edge number of a graph G, denoted by η(G), is the smallest natural number p selected from the set {1, 2, ..., p}.

Sum Labeling

A graph is classified as a sum graph if there exists a one-to-one function ϕ from its vertex set V to a collection of natural numbers N, such that for any edge $s_1s_2 \in E$, the condition $\phi(s_1) + \phi(s_2) \in N$ is satisfied. Additionally, the sum of the labels of vertices connected by an edge must be equal to the label of any other vertex in the graph. It is imperative for sum graphs to have isolated vertices because the vertex with the highest label cannot be in proximity to any other vertex.

Inclusive Lucky labeling

Consider a vertex labeling, denoted by *l*: V(G) → N, where each vertex in the graph G is assigned a positive integer. For two adjacent vertices, v_1 and v_2, the labeling *l* is termed an inclusive lucky labeling if the sum S[v1] is not equal to the sum S[v2]. The expression for $S[v_1]$ is defined as $S[v_1] = \sum_{vi \in N(v1)} l(vi) + l(v1)$. For an isolated vertex v1, the label $l(v_1)$ is set to 1, and consequently, $S[v_1]$ equals 1. The inclusive lucky number associated with each inclusive lucky labeling is denoted as η (G). In the context of a graph G, the inclusive lucky number represents the smallest positive integer p, given that the labels assigned to its vertices belong to the set {1, 2, 3, 4, ..., p}.

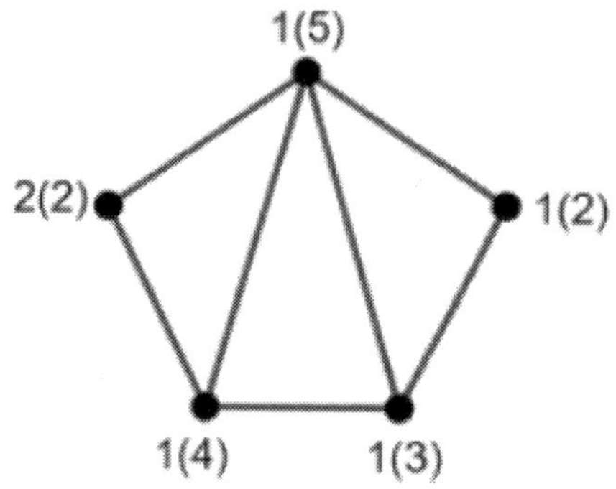

Proper Inclusive Lucky Labeling

Let $l : V(G) \to N$ represent a vertex labeling where each vertex in a graph G is as- signed a positive integer. For any pair of adjacent vertices v_1 and v_2, define labeling l as proper inclusive lucky labeling if $S[v_1] = S[v_2]$ and $l(v_1) = l(v_2)$; $S[v_1]$ is defined as $S[v_1] = \sum_{v_i \in N(v_1)} l(v_i) + l(v_1)$. The inclusive lucky number for any inclusive lucky label-

ing is $\eta_{ip}(G)$. When a graph G has vertices label sets of $\{1,2,3,4,\ldots,p\}$, its smallest positive integer p is considered its proper inclusive lucky number.

d-Lucky Labeling

Consider a vertex labeling denoted by d: V(G) → N, where each vertex in the graph G is assigned a positive integer. If the condition $c(s_1) \neq c(s_2)$ holds for all pairs of adjacent vertices s_1 and s_2, then the labeling d is defined as a d-lucky labeling. The mathematical representation of c(v) is given by $c(v) = \sum_{vi \in N(v)} d(vi) + deg(v)$. The d-lucky number associated with each d-lucky labeling is denoted as ηdl(G). In the case where a graph G has vertices labeled from the set {1, 2, 3, 4, ..., p}, the smallest positive integer p is regarded as its d-lucky number.

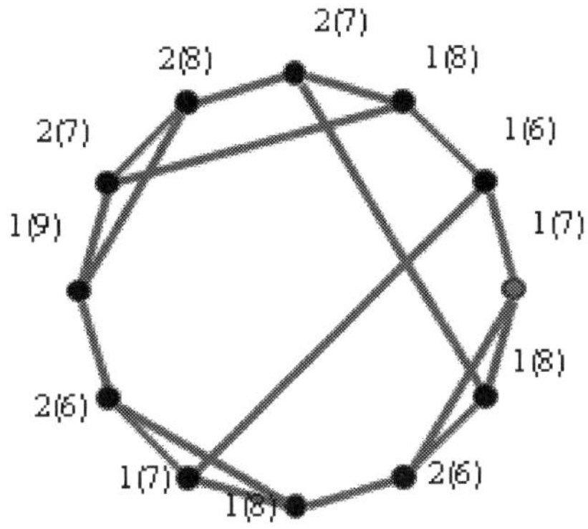

L(2,1) Labeling

A function f from $V(G)$ to nonnegative integers is said to be $L(2,1)$ labeling if it satisfies the following conditions:

(1) $|f(u) - f(v)| \geq 2$, $if\ d(u,v) = 1$

(2) $|f(u) - f(v)| \geq 1$, $if\ d(u,v) = 2$

where $d(u,v)$ is the distance between vertices u and v.

Remark: The $L(2,1)$-labeling number $\lambda(G)$, also known as span of labeling or λ-number, is the smallest number l such that $l = \max\{f(v) | v \in V(G)\}$ among all L(2,1) labelings.

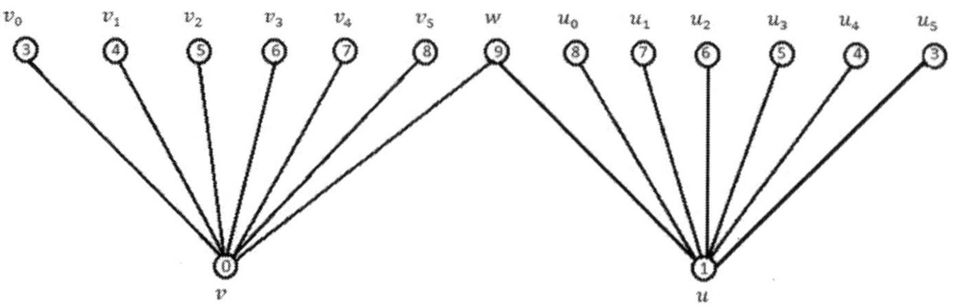

Definition 2.3 An olive tree T_n is a rooted tree having n branches, where the kth branch is considered as a path of length k. The number of vertices of olive tree is $|V(T_n)| = \{1 + 2 + \ldots(n-2) + (n-1) + n\} + 1$ and number of edges is $|E(T_n)| = V(T_n) - 1$

Definition 2.4 Let p_n^1 and p_n^2 be the parallel paths having n vertices and let $v_o, v_1, v_2, \ldots\ldots\ldots, v_{n-1}$ be the vertices of p_n^1 also $u_o, u_1, u_2, \ldots\ldots\ldots, u_{n-1}$ be the vertices of p_n^2. The H-graph is obtained by joining p_n^1 and p_n^2 with an edge. If n is odd, the edge joins the vertices $v_{\frac{n+1}{2}}$ and $u_{\frac{n+1}{2}}$, if n is odd then edge joins the vertices $v_{\frac{n}{2}}$ and $u_{\frac{n}{2}+1}$. It is denoted by H_n. The number of vertices of H-graph is $|V(H_n)| = 2n$ and number of edges is $|E(H_n)| = 2(n-1) + 1$

Definition 2.5 Let $k_{1,n}^1$ and $k_{1,n}^2$ be the two copies of star graphs. The W-graph is obtained by merging the last pendant vertex of $k_{1,n}^1$ with first pendant vertex of $k_{1,n}^2$. It is denoted by W_n. The number of vertices of W-graph is $|V(W_n)| = 2n + 1$ and number of edges is $|E(W_n)| = 2n$.

3. Main Result

Theorem 3.1 The λ- number of olive tree T_n is $n + 1$.

Proof. Let v be a root of olive tree T_n and B_1, B_2, B_3, ..., B_k be the branches of olive tree graph.

B_1 be the first branch having one vertex,

B_2 be the second branch having two vertices,

.

.

.

B_k be the kth branch having k vertices.

Let u_0^k, v_0^k, v_1^k, ..., v_{k-2}^k be the vertices of branch B_k and u_0^k be the vertices directly connected to root v, here $k = 1, 2, 3,..., n$.

Now define $f : V(T_n) \rightarrow \{0, 1, 2, 3, ..., n + 1\}$ as follows:

$f(v) = 0$

$$f(v_i^k) = \begin{cases} 1 & , i = 0 \; modulo \; 3 \\ 3 & , i = 1 \; modulo \; 3 \\ 5 & , i = 2 \; modulo \; 3 \end{cases}$$

where $k = 1, 2, 3, ..., n$

and

$$f(u_0^k) = \begin{cases} 2 & , \; if \; k = 1 \\ 3 & , \; if \; k = 2 \\ 4 & , \; if \; k = 3 \\ . & , \quad . \\ . & , \quad . \\ . & , \quad . \\ n+1 & , \; if \; k = n \end{cases}$$

Hence $\lambda(T_n) = n + 1$.

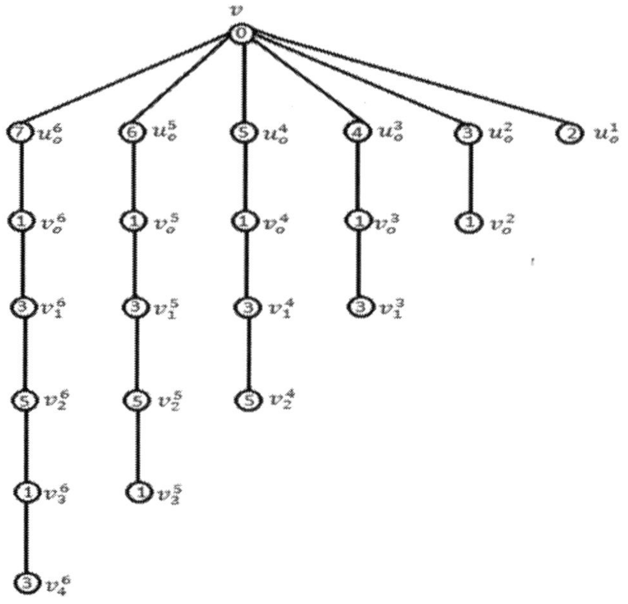

Figure 1: $L(2,1)$ labeling of olive tree T_6 with labeling number 7.

Theorem 3.2: The λ- number of H-graph H_n is

$$\lambda(H_n) = \begin{cases} 3 & if \ n = 2 \\ 5 & if \ n \geq 3 \end{cases}$$

Proof: Let p_n^1 and p_n^2 be the parallel paths of H-graph H_n and let v_o , v_1 , v_2 , ... , v_{n-1} be the vertices of p_n^1 in H-graph H_n such that v_i is adjacent to v_{i+1} and let u_o , u_1 , u_2 , ... , u_{n-1} be the vertices of p_n^2 in H-graph such that u_i is adjacent to u_{i+1} also v_l is adjacent to u_l and when n is odd then $v_{\frac{n+1}{2}}$ and $u_{\frac{n+1}{2}}$ are adjacent but when n is even then $v_{\frac{n}{2}}$ and $u_{\frac{n}{2}+1}$ are adjacent. The L (2,1)-labeling number of H_2 is 3 because H_2 is isomorphic to P_4 and have labels 2,0,3,1 respectively.

Now define

$f : V(H_n) \rightarrow \{0, 1, 2, 3,4,5\}$ as:

Case 1. If n is even

$$f(v_j) = \begin{cases} 5 & , j = 3\varrho \\ 3 & , j = 3\varrho + 1 \\ 1 & , j = 3\varrho + 2 \end{cases}$$

and

$$f(u_i) = \begin{cases} 4 & , i = 3\varrho \\ 2 & , i = 3\varrho + 1 \\ 0 & , i = 3\varrho + 2 \end{cases}$$

Case 2. If n is odd

$$f(v_j) = \begin{cases} 5 & , j = 3\varrho \\ 3 & , j = 3\varrho + 1 \\ 1 & , j = 3\varrho + 2 \end{cases}$$

and

$$f(u_i) = \begin{cases} 2 & , i = 3\varrho \\ 0 & , i = 3\varrho + 1 \\ 4 & , i = 3\varrho + 2 \end{cases}$$

Here, s is a non-negative integer.

Hence, labeling number of H_n is

$$\lambda(H_n) = \begin{cases} 3 & if \;\; n = 2 \\ 5 & if \;\; n \geq 3 \end{cases}$$

Example 3.3

In figure 3, the $L(2,1)$ labeling of H-graph H_8 is shown.

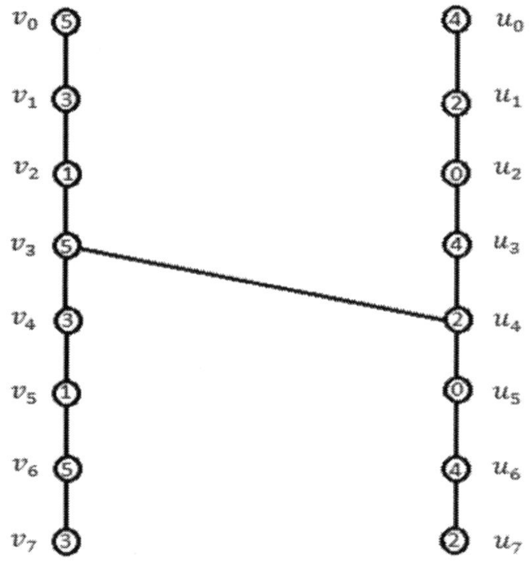

Figure 3

Example 3.4

In figure 4, the $L(2,1)$ labeling of H-graph H_7 is shown.

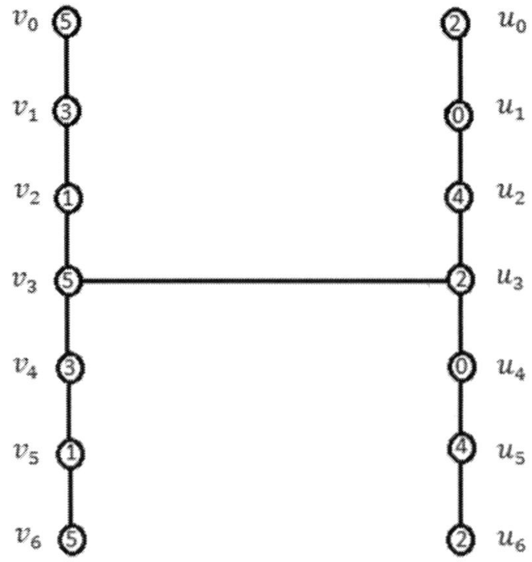

Figure 4

Theorem 3.3: The λ- number of W-graph W_n is

$$\lambda(W_n) = \begin{cases} 4 & if\ n = 2 \\ n+2 & if\ n \geq 3 \end{cases}$$

Proof: Let $k_{1,n}^1$ and $k_{1,n}^2$ be the two star graphs. Let v be the root of $k_{1,n}^1$ and v_o , v_1 , v_2 , ..., v_{n-2} , w_1 be the vertices of $k_{1,n}^1$ such that these are directly connected to root v Also u be the root of $k_{1,n}^2$ and w_2 ,u_o , u_1 , u_2 , ..., u_{n-2} be the vertices of $k_{1,n}^2$ such that these vertices are directly connected to root u . We obtained the W-graph W_n by joining the w_1 and w_2 having named w and w is directly connected to v and u . The $L\ (2,1)$-labeling number of W_2 is 4 because W_2 is isomorphic to P_5 and have labels $4,2,0,3,1$ respectively.

Now define

$f : V(W_n) \rightarrow \{0, 1, 2, 3, ..., n+2\}$ as follows:

$f(v) = 0$

$f(u) = 1$

$$f(v_i) = \begin{cases} 3 & , if\ i = 0 \\ 4 & , if\ i = 1 \\ 5 & , if\ i = 2 \\ 6 & , if\ i = 3 \\ \cdot & \\ \cdot & \\ \cdot & \\ n & ,if\ i = n-1 \\ n+1 & , if\ i = n-2 \end{cases}$$

$$f(u_i) = \begin{cases} n+1 & , if\ i = 1 \\ n & , if\ i = 2 \\ n-1 & , if\ i = 3 \\ n-2 & , if\ i = 4 \\ \cdot & \\ \cdot & \\ \cdot & \\ 2 & ,if\ i = n-1 \\ 3 & , if\ i = n-2 \end{cases}$$

$f(w) = n + 2$

Hence, labeling number of W-graph is

$$\lambda(W_n) = \begin{cases} 4 & if\ n = 2 \\ n + 2 & if\ n \geq 3 \end{cases}$$

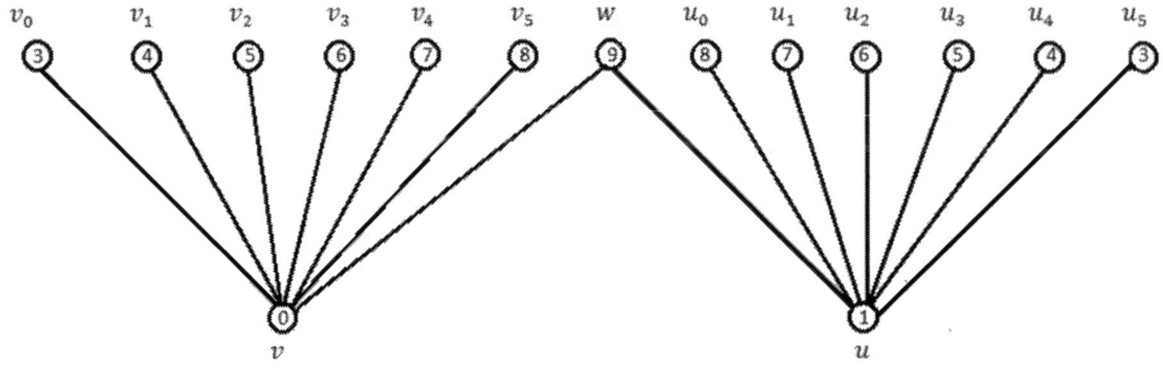

Figure 5: $L(2,1)$ labeling of W-graph W_7.

Chapter 7

Irregularity of Wheel Derived graphs

For any connected graph G the irregularity of graph is define as the sum of imbalance of a line (edge) $e = uv$ such that $irr(G) = \sum_{uv \in E(G)} imb(e) = \sum_{uv \in E(G)} |d_G(v) - d_G(u)|$, where imbalance of an edge is defined as $imb(e) = |d_G(v) - d_G(u)|$. A new measurement of irregularity known as total irregularity of a graph which is described as $irr_t(G) = \frac{1}{2} \sum_{u,v \in V(G)} |d_G(v) - d_G(u)|$ will also discussed. For any connected graph variance of degree defined as $Var(G) = \frac{1}{n} \sum_{i=1}^{n} d_i^2 - \frac{1}{n^2} (\sum_{i=1}^{n} d_i)^2$. In order to calculate all of these irregularities, edge and vertex partition will be calculated and with the help of these partitions these three types of irregularities will be calculated. For given graph total irregularity is completely determined by graphic sequence. This sequence is state as a sequence of zero and positive integers $d_1, \ldots, d_n$. Graphs have same total irregularities if their degree sequence is same but in case of regularity, this is not always true.

Wheel graph

A wheel graph is denoted as W_n which is obtained from a cycle so that each of its vertex is connected by a single vertex. Vertex set and edges set of wheel graph is defined as

$V(G) = \{t, t_i, 1 \leq i \leq n\}$

$E(G) = \{tt_i, t_i t_{i+1}, 1 \leq i \leq n\}$

In order to calculate irregularity by Albertson, total irregularity and variance of degree of wheel graph first of all there is a need to generalize vertex and edge partitions of given graph. Now we will analysis the structure of wheel graph.

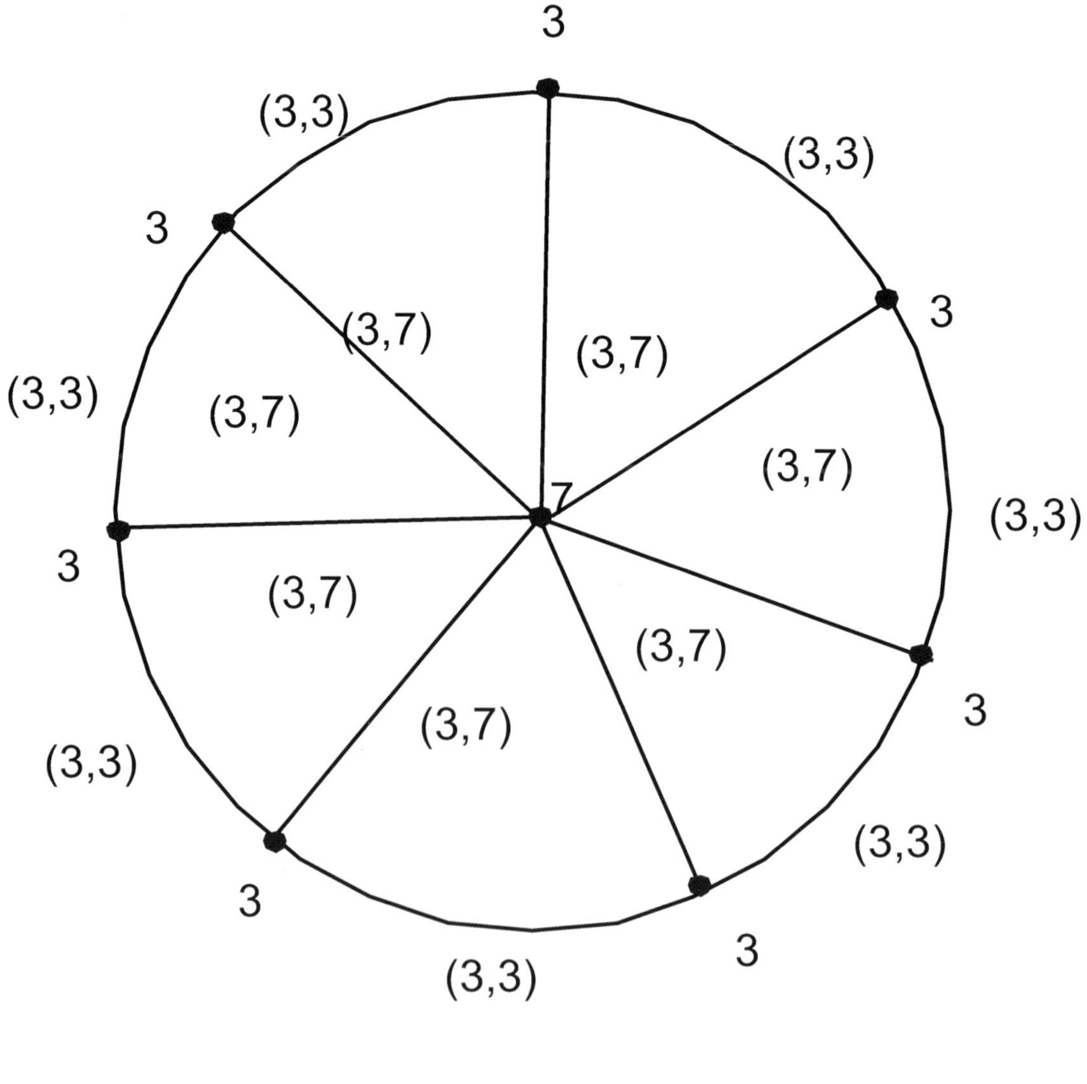

Figure 3.1: A wheel graph W_7.

After this example the general vertex partition of wheel graph is given as

Table 1. Vertex partition of Wheel graph.

W_n	
d_u	No. of vertices
3	n
n	1

Edge partition of wheel graph is given as

Table 2. Edge partition of Wheel graph.

W_n	
(d_u, d_v)	Number of edges
(3 , 3)	n
(n , 3)	n

Fan graph

This graph is symbolized as F_n which can be generated by a wheel graph in such a way that one edge is deleted from cycle C_n . Vertex set and edges set of fan graph is defined as

$V(G) = \{t, t_i \ , 1 \leq i \leq n\}$

$E(G) = \{tt_i \, , t_i t_{i+1} \, , 1 \leq i \leq n - 1\}$

In order to calculate irregularity by Albertson, total irregularity and variance of degree of fan graph first of all there is a need to generalize vertex and edge partitions of given graph. Now we will analysis the structure of fan graph.

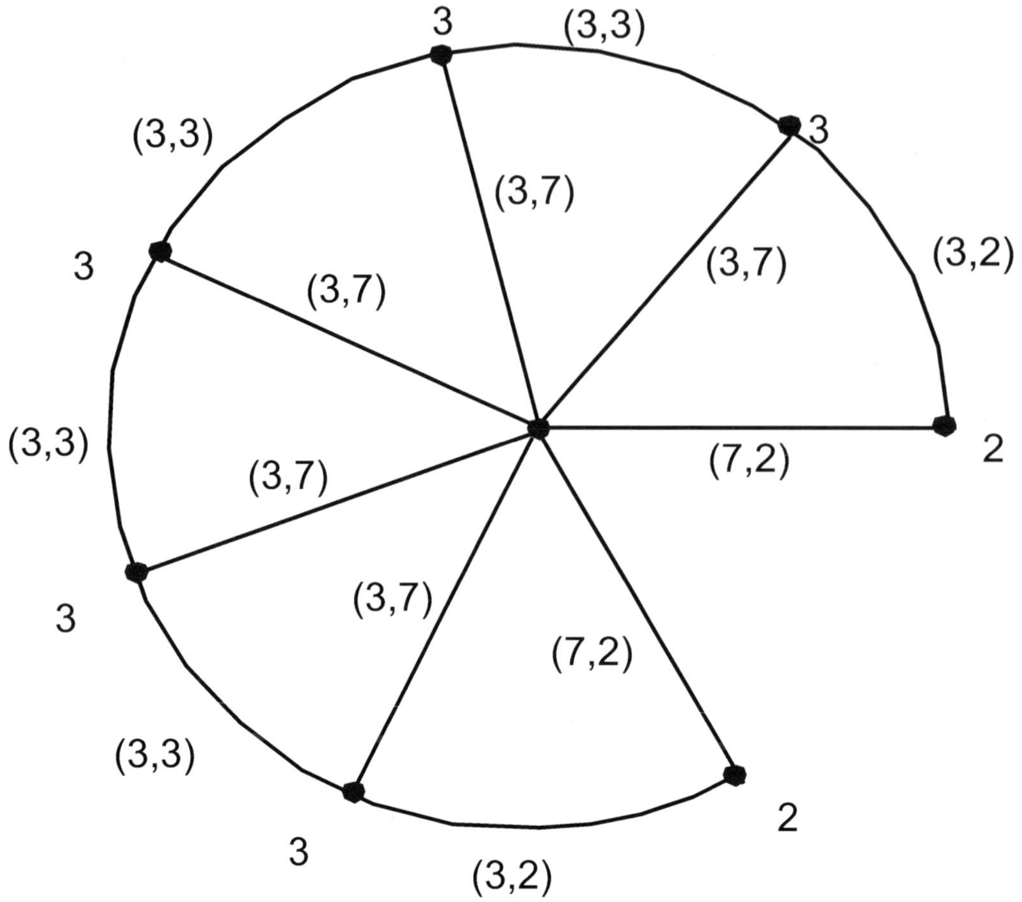

Figure 3.2: A fan graph F_7.

After this the general vertex partition of fan graph is given as

Table 3. Vertex partition of Fan graph.

F_n	
d_u	Number of vertex
2	2
3	n-2
n	1

Edge partition of fan graph is given as

F_n	
(d_u, d_v)	Number of edges
(3 , 3)	(n-3)
(n,3)	(n-2)
(n,2)	2
(3,2)	2

Helm graph

This graph is symbolized as H_n which is generated by wheel graph such that all of its vertex in C_n one pendent edge is added. Vertex set and edges set of helm graph is defined as

$V(G) = \{t, t_i, s_i, 1 \leq i \leq n\}$

$E(G) = \{tt_i, t_it_{i+1}, t_is_i, 1 \leq i \leq n\}$

Indices taken modulo n.

In order to calculate irregularity by Albertson, total irregularity and variance of degree of helm graph first of all there is a need to generalize vertex and edge partitions of given graph. Now we will analysis the structure of helm graph.

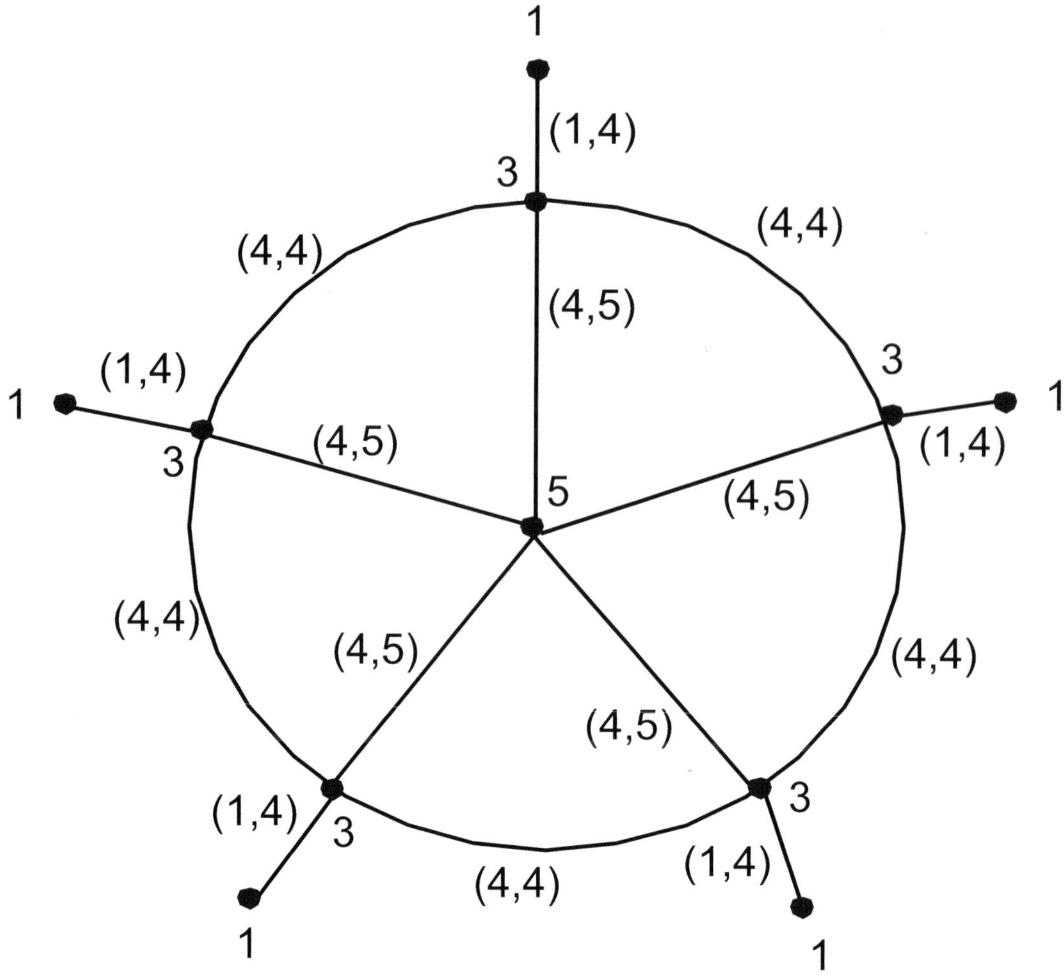

Figure 3.3: A helm graph H_5.

After this example the general vertex partition of helm graph is given as

Table 5. Vertex partition of Helm graph.

H_n	
d_u	Number of vertex
1	n
4	n
n	1

Edge partition of helm graph is given as

Table 6. Edge partition of helm graph.

H_n	
(d_u, d_v)	Number of edges
(4 , 1)	n
(4 , 4)	n
(n , 4)	n

Web graph

A web graph is denoted as Wb_n which is obtained from helm graph such that there is an edge incident between each pair of pendent vertex to form a cycle. Vertex set and edges set of web graph is define as

$V(G) = \{t, t_i, s_i, r_i, 1 \leq i \leq n\}$

$E(G) = \{tt_i, t_i t_{i+1}, t_i s_i, s_i s_{i+1}, s_i r_i \; 1 \leq i \leq n\}$

Indices taken modulo n.

In order to calculate irregularity by Albertson, total irregularity and variance of degree of web graph first of all there is a need to generalize vertex and edge partitions of given graph. Now we will analysis the structure of web graph.

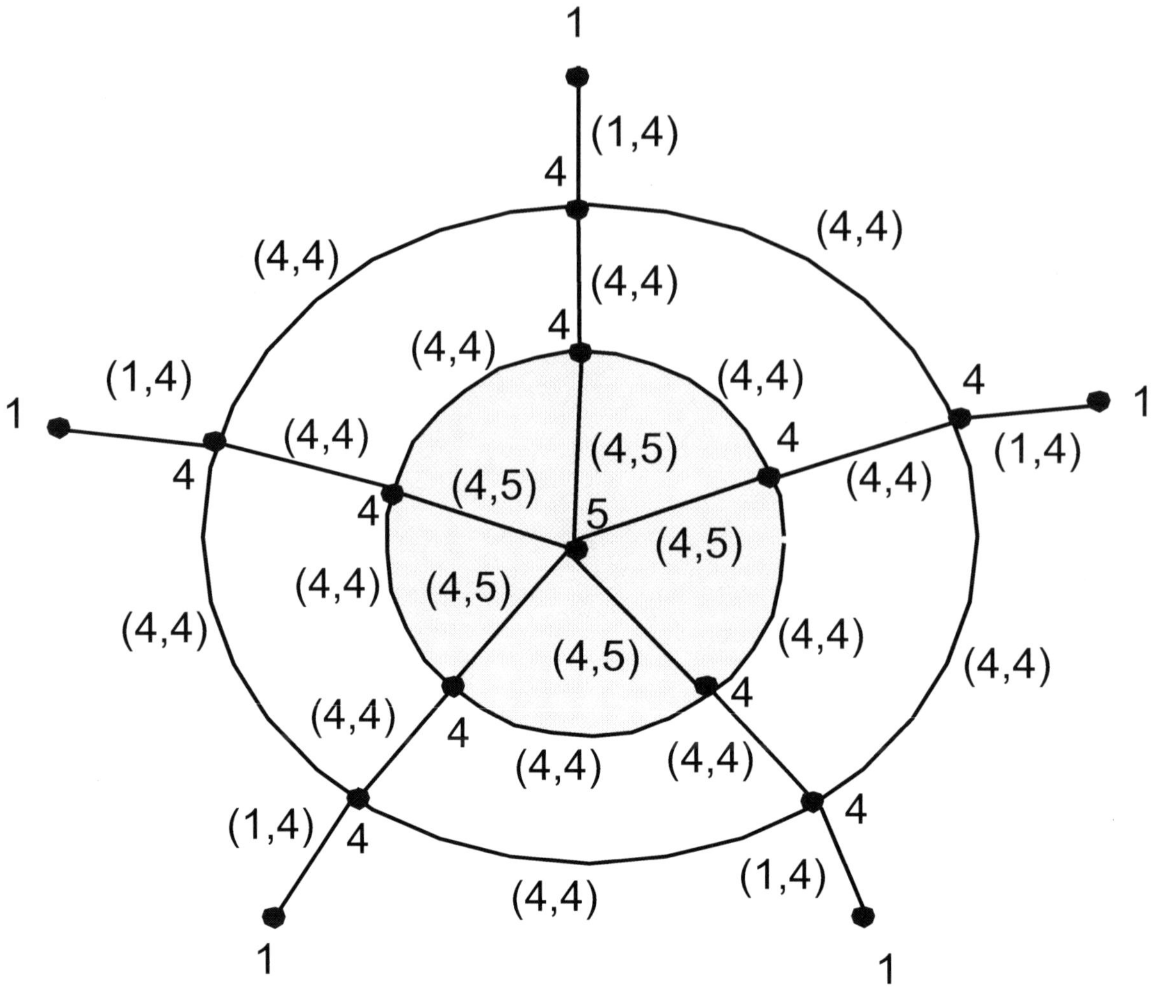

Figure 3.4: A web graph Wb_5

After this the general vertex partition of web graph is given as

Table 7. Vertex partition of Web graph.

Wb_n	
d_u	Number of vertex
1	n
4	2n
n	1

Edge partition of web graph is given as

Table 8. Edge partition of Web graph

Wb_n	
(d_u, d_v)	Number of edges
(4 , 1)	n
(4 , 4)	3n
(n , 4)	n

Flower graph

A flower graph is denoted as Fl_n which is obtained from helm graph in such a way that each pendant vertex is joined by central vertex of helm graph. Vertex set and edges set of flower graph is define as

$V(G) = \{t, t_i, s_i, 1 \leq i \leq n\}$

$E(G) = \{tt_i, ts_i, t_it_{i+1}, t_is_i, 1 \leq i \leq n\}$

Indices taken modulo n.

In order to calculate irregularity by Albertson, total irregularity and variance of degree of flower graph first of all there is a need to generalize vertex and edge partitions of given graph. Now we will analysis the structure of flower graph.

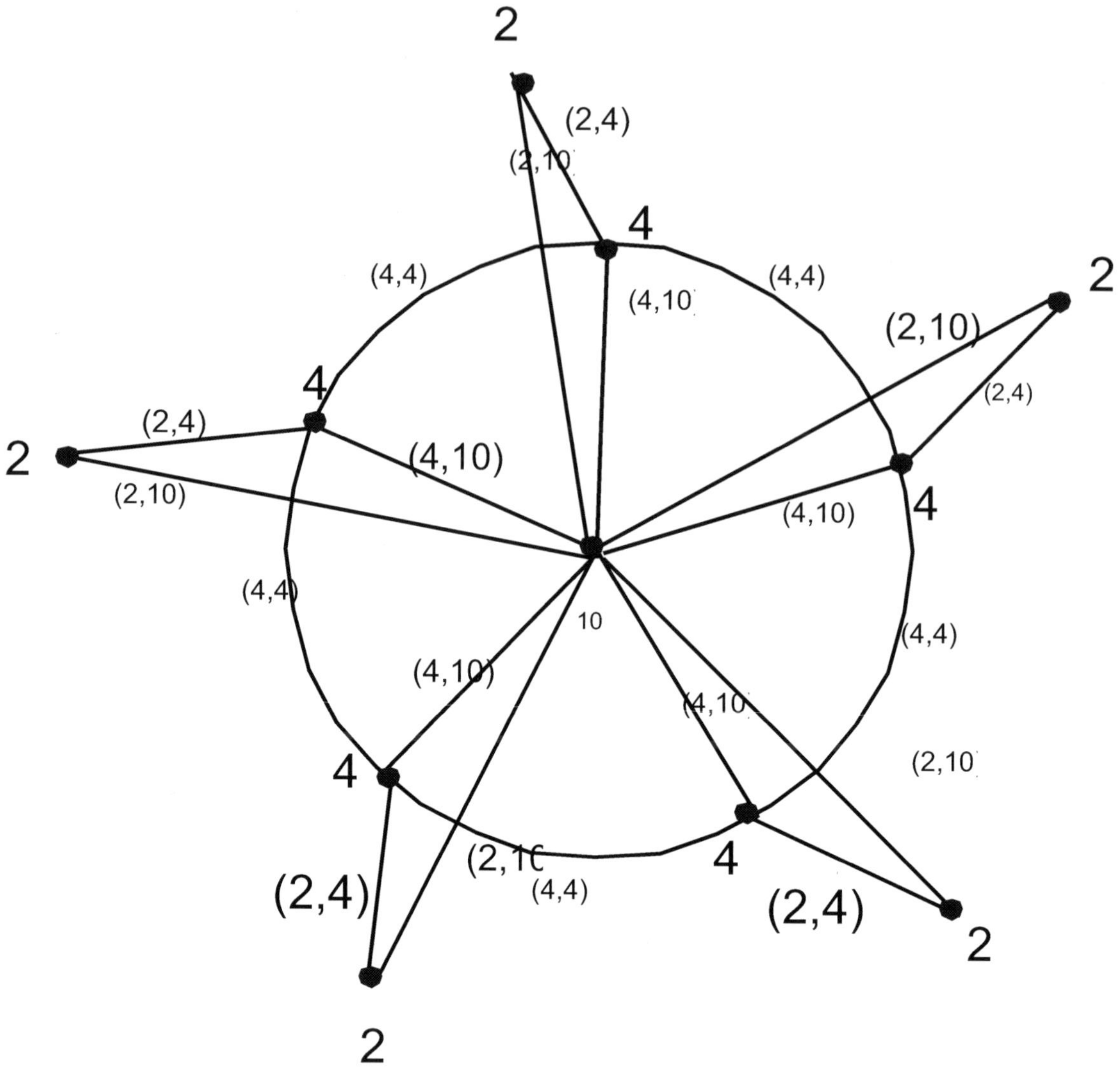

Figure 3.5: A flower graph Fl_5

After this the general vertex partition of flower graph is given as

Table 9. Vertex partition of Flower graph.

Fl_n	
d_u	Number of vertex
2	n
4	n
2n	1

Edge partition of flower graph

Table 10. Edge partition of Flower graph.

Fl_n	
(d_u, d_v)	Number of edges
(2n , 2)	n
(4 , 2)	n
(4 , 4)	n
(2n , 4)	n

Sunflower graph

A sunflower graph is denoted as SF_n which is obtained from flower graph in such a way that n pendant edges are added to the central vertex. Vertex set and edges set of sunflower graph is define as:

$V(G) = \{t, t_i, s_i, r_i, 1 \leq i \leq n\}$

$E(G) = \{tt_i, ts_i, tr_i, t_it_{i+1}, t_is_i, 1 \leq i \leq n\}$

Indices taken modulo n.

In order to calculate irregularity by Albertson, total irregularity and variance of degree of sunflower graph first of all there is a need to generalize vertex and edge partitions of given graph. Now we will analysis the structure of sunflower graph.

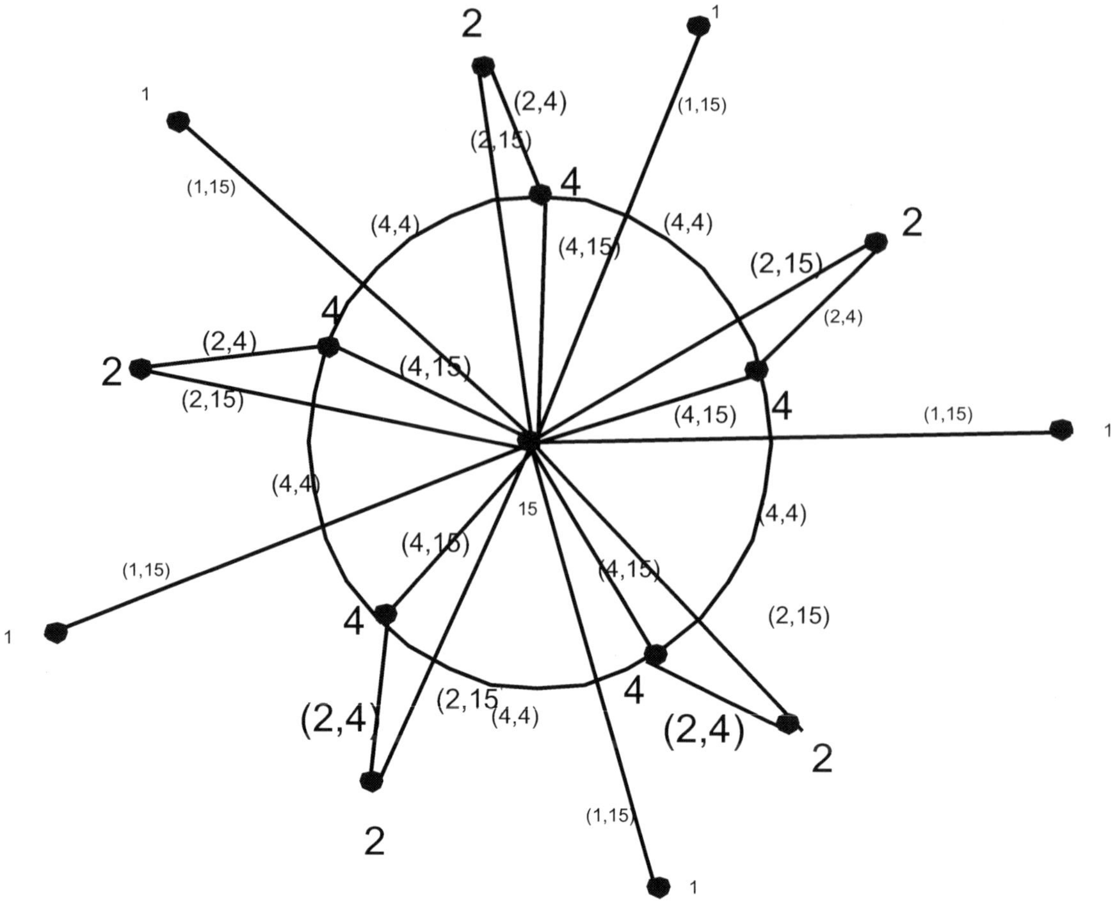

Figure 3.6: A sun flower graph SF_5

After this the general vertex partition of sun flower graph is given as

Table 11. Vertex partition of sun flower graph.

SF_n	
d_u	Number of vertex
2	n
4	n
3n	1
1	n

Edge partition of sun flower graph is given as

Table 12. Edge partition of sun flower graph.

SF_n	
(d_u, d_v)	Number of edges
(3n , 2)	n
(4 , 2)	n
(4 , 4)	n
(3n , 4)	n
(3n , 1)	n

Gear graph

A gear graph is denoted as G_n which is obtained from wheel graph in such a way that subdivide the vertices of outer cycle. Vertex set and edges set of gear graph is define as:

$V(G) = \{t, t_i, s_i, 1 \leq i \leq n\}$

$E(G) = \{tt_i, t_is_i, s_it_{i+1}, 1 \leq i \leq n\}$

Indices taken modulo n.

In order to calculate irregularity by Albertson, total irregularity and variance of degree of gear graph first of all there is a need to generalize vertex and edge partitions of given graph. Now we will analysis the structure of gear graph.

.

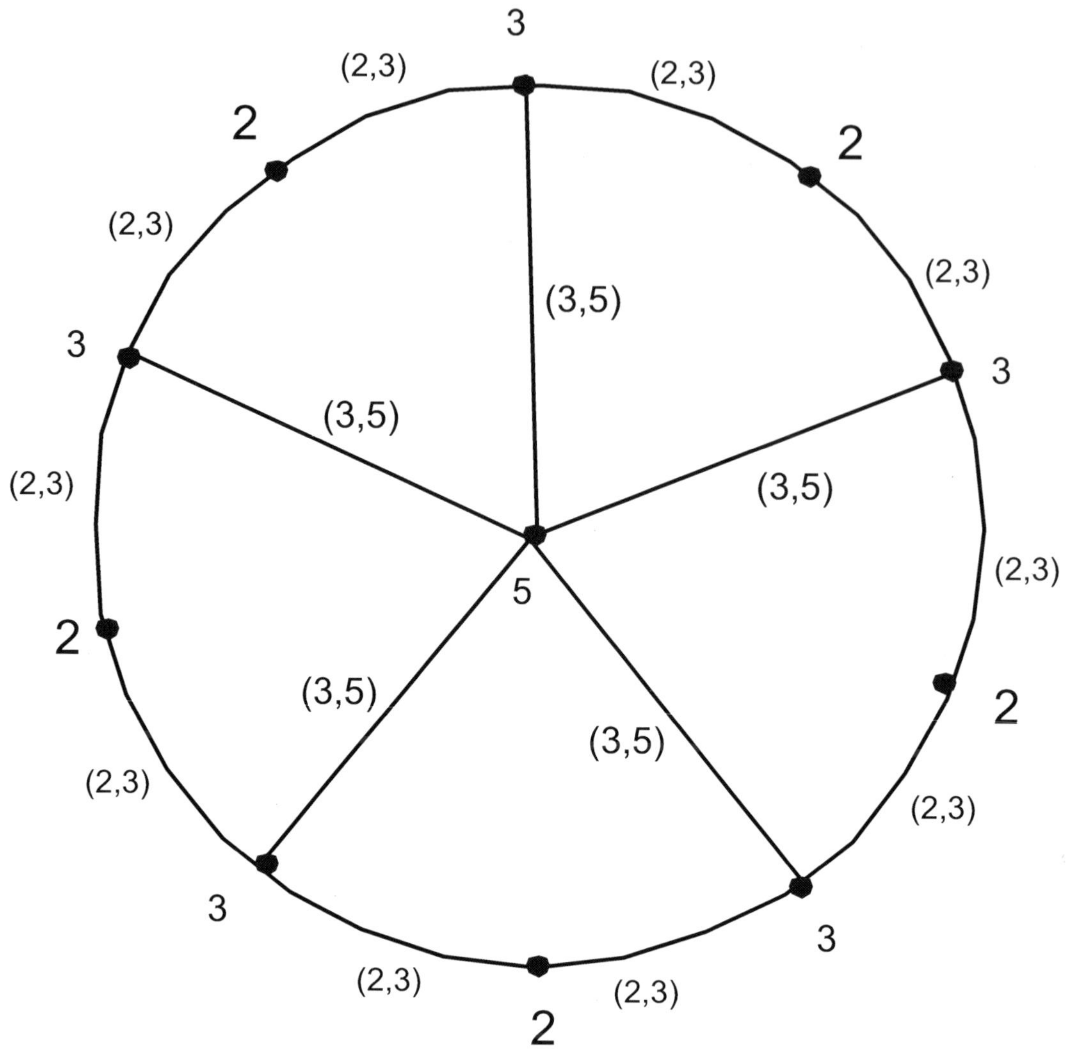

Figure 3.7: A gear graph G_5

After this the general vertex partition of gear graph is given as

Table 13. Vertex partition of gear graph.

G_n	
d_u	Number of vertex
3	n
2	n
n	1

Edge partition of gear graph is given as

Table 14. Edge partition of gear graph.

G_n	
(d_u, d_v)	Number of edges
(3 , 2)	2n
(n , 3)	n

Closed helm graph

This graph is symbolized as CH_n which is generated by elm graph in such a way that each pendant vertex of helm graph are joined to form a cycle. Vertex set and edges set of closed helm graph is define as:

$V(G) = \{t, t_i, s_i, 1 \leq i \leq n\}$

$E(G) = \{tt_i, ts_i, t_it_{i+1}, s_is_{i+1}, 1 \leq i \leq n\}$

Indices taken modulo n.

In order to calculate irregularity by Albertson, total irregularity and variance of degree of closed helm graph first of all there is a need to generalize vertex and edge partitions of given graph. Now we will analysis the structure of closed helm graph.

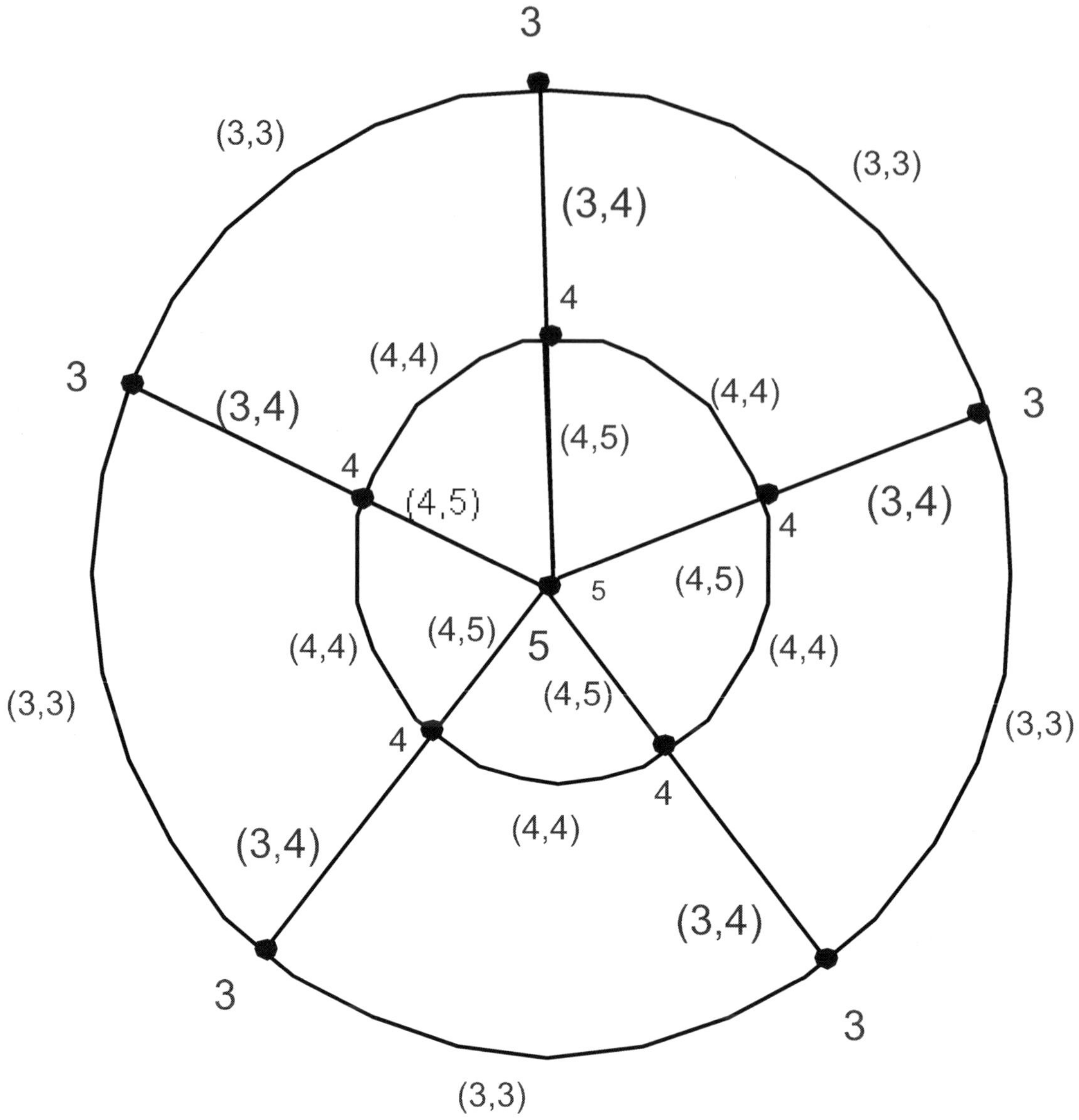

Figure 3.8: A closed helm graph CH_5

After this the general vertex partition of closed helm graph is given as

Table 15. Vertex partition of closed helm graph.

CH_n	
d_u	Number of vertex
3	n
4	n
n	1

Edge partition of closed helm graph is given as

Table 16. Edge partition of closed helm graph.

CH_n	
(d_u, d_v)	Number of edges
(3 , 4)	n
(3 , 3)	n
(4 , 4)	n
(n , 4)	n

Dutch Windmill graph

A Dutch windmill graph is denoted as D_n^m which is obtained from cycle graph C_n in such a way that m copies of C_n are attached with one common vertex. In order to calculate irregularity, total irregularity and variance of degree of Dutch windmill graph first of all there is a need to generalize vertex and edge partitions of given graph. Now we will analysis the structure of closed helm graph.as

Friendship graph

A friendship graph is a Dutch win mill graph which is denoted as D_3^m which is obtained from cycle of length 3 (C_3) in such a way that m copies of C_3 are attached with one common vertex.

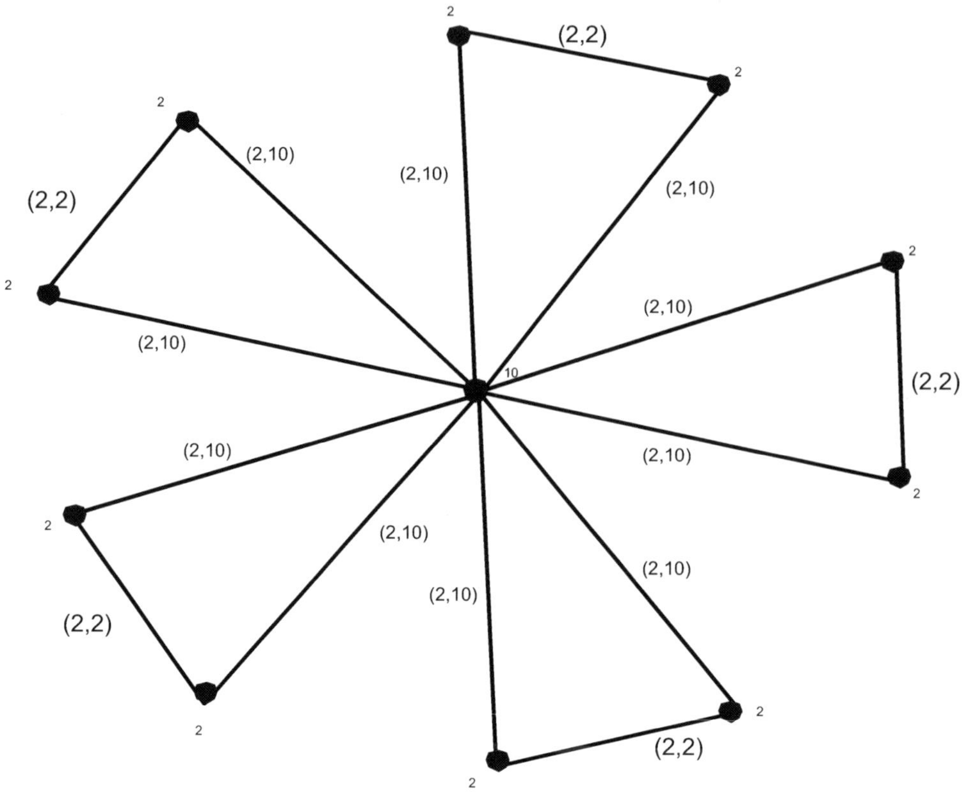

Figure 3.9: A friendship graph D_3^m

From above figure we can easily generalize the vertex and edge partition of D_3^m graph . The edge partition of D_3^m graph is given as

Table 17. Edge partition of friendship D_3^m graph.

D_3^m	
(d_u, d_v)	Number of edges
(2 , 2)	m
(2 , 2m)	2m

The vertex partition of D_3^m graph is given as

Table 18. Vertex partition of friendship D_3^m graph

D_3^m	
d_u	Number of vertex
2	2m
2m	1

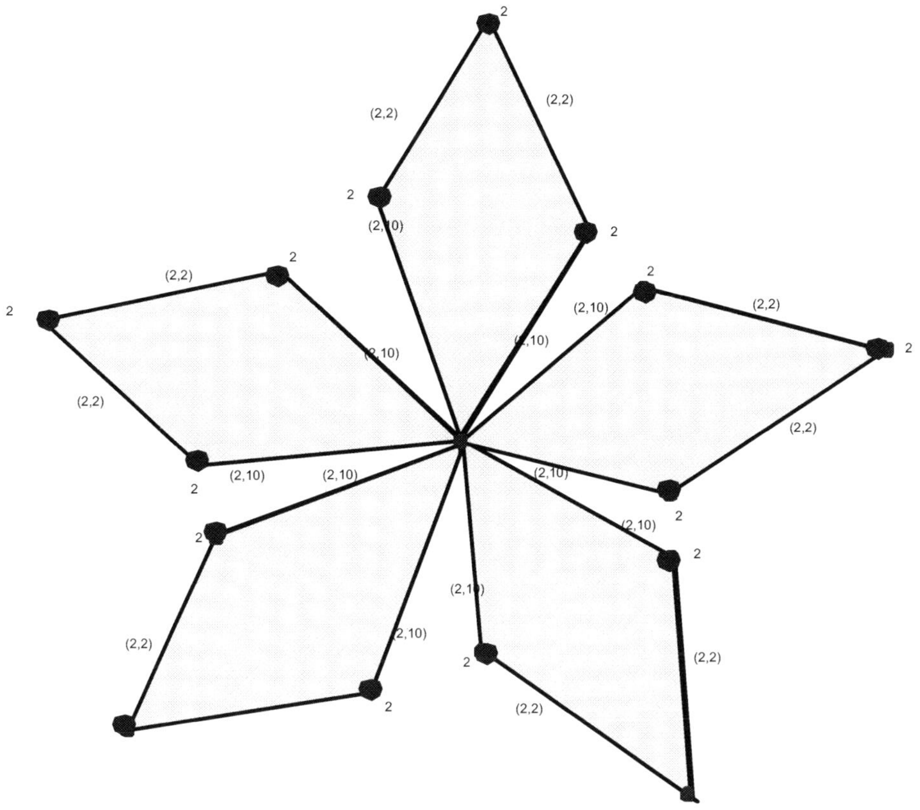

Figure 3.10: A Dutch windmill graph D_4^m

From above figure we can easily generalize the vertex and edge partition of D_4^m graph.

The edge partition of D_4^m graph is given as

Table 19. Edge partition of D_4^m graph.

D_4^m	
$d_u,d_v)$	Number of edges
(2 , 2)	2m
(2 , 2m)	2m

The vertex partition of D_4^m graph is given as

Table 20. Vertex partition of D_4^m graph.

D_4^m	
d_u	Number of vertex
2	3m
2m	1

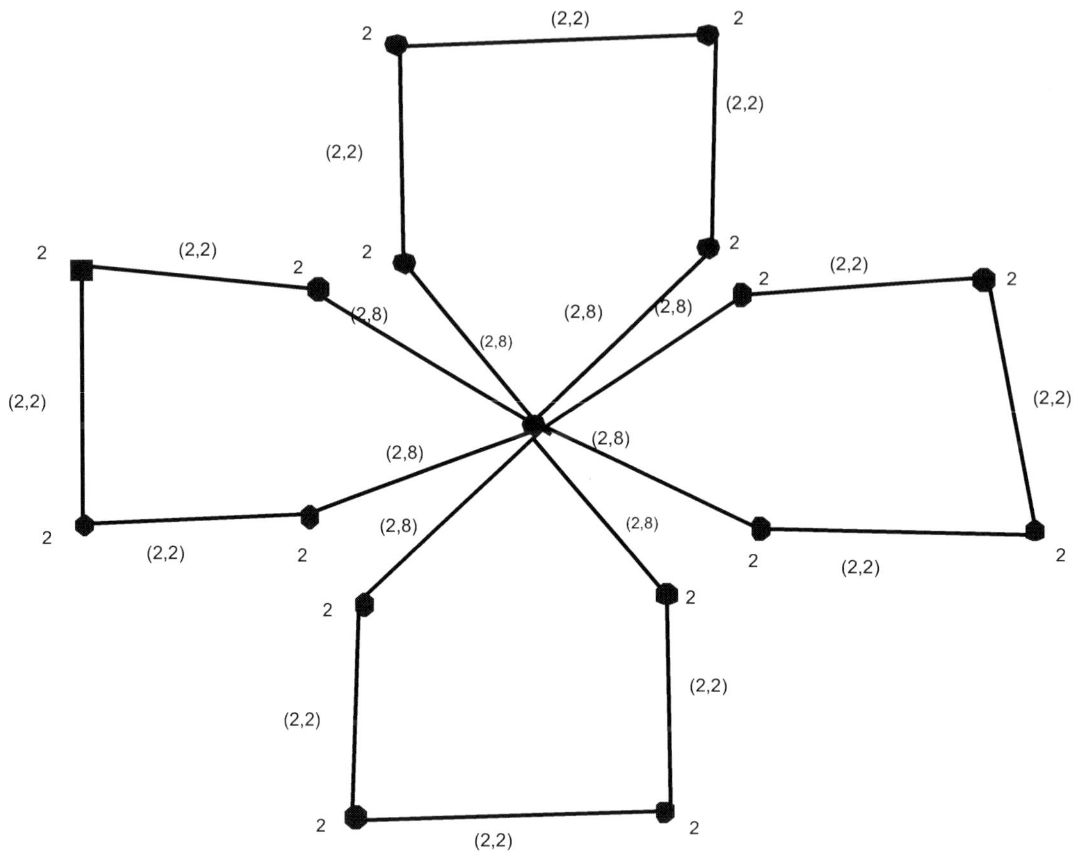

Figure 3.11: A Dutch windmill graph D_5^m

From above figure we can easily generalize the vertex and edge partition of D_4^mgraph.

The edge partition of D_5^m graph is given as

Table 21. Edge partition of D_5^m graph.

D_5^m	
$d_u,d_v)$	Number of edges
(2 , 2)	3m
(2 , 2m)	2m

The vertex partition of D_5^m graph is given as

Table 22. Vertex partition of D_5^m graph.

D_5^m	
d_u	Number of vertex
2	4m
2m	1

From all of above example now we can easily generalize edge partition of D_n^m Dutch windmill graph.

Table 23. Edge partition of D_n^m graph.

D_n^m	
(d_u, d_v)	Number of edges
(2 , 2)	(n-2)m
(2 , 2m)	2m

From all of above example now we can easily generalized vertex partition of D_n^m Dutch windmill graph.

Table 24. Vertex partition of D_n^m graph.

D_n^m	
d_u	Number of vertices
2	(n-1)m
2m	1

Windmill graph

A wind mill graph is denoted as W_n^m which is obtained from complete graph K_n in such a way that m copies of K_n are attached with one common vertex. In order to calculate irregularity, total irregularity and variance of degree of wind mill graph first of all there is a need to generalized vertex and edge partitions of given graph. For this purpose a deep study on structure of this graph is required.

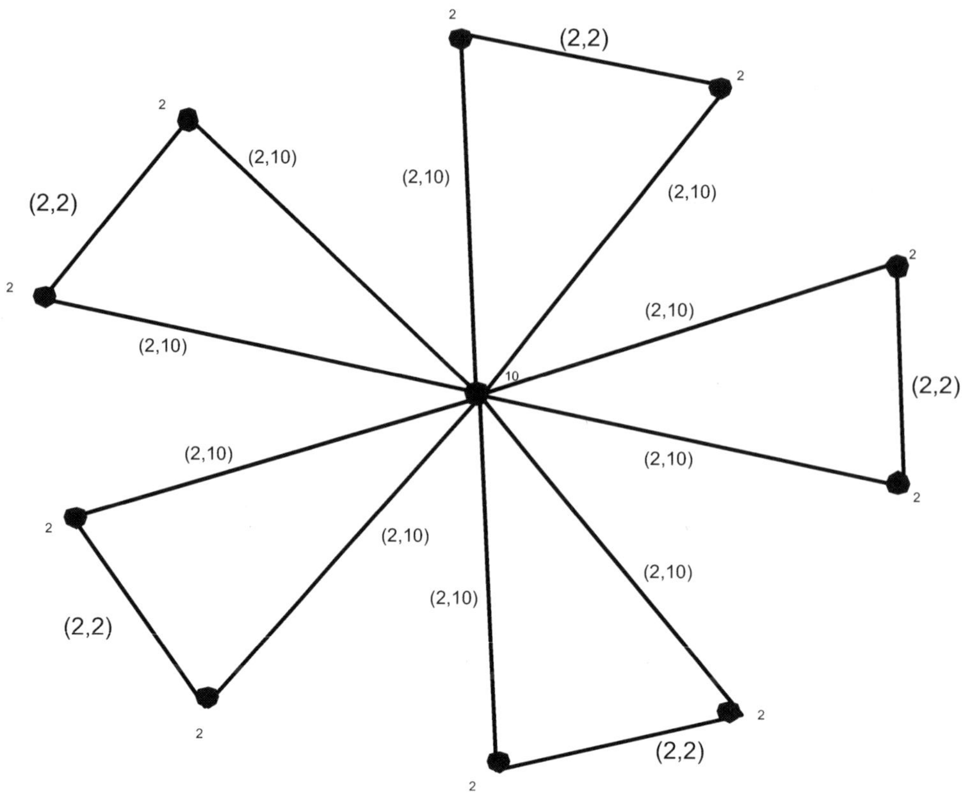

Figure 3.12: A windmill graph W_3^m

From above figure we can easily generalize the vertex and edge partition of W_3^m graph .

The edge partition of W_3^m graph is given as

Table 25. Edge partition of W_3^m graph.

W_3^m	
(d_u, d_v)	Number of edges
(2 , 2)	m
(2 , 2m)	2m

The vertex partition of W_3^m graph is given as

Table 26. Vertex partition of W_3^m graph.

W_3^m

d_u	Number of vertex
2	2m
2m	1

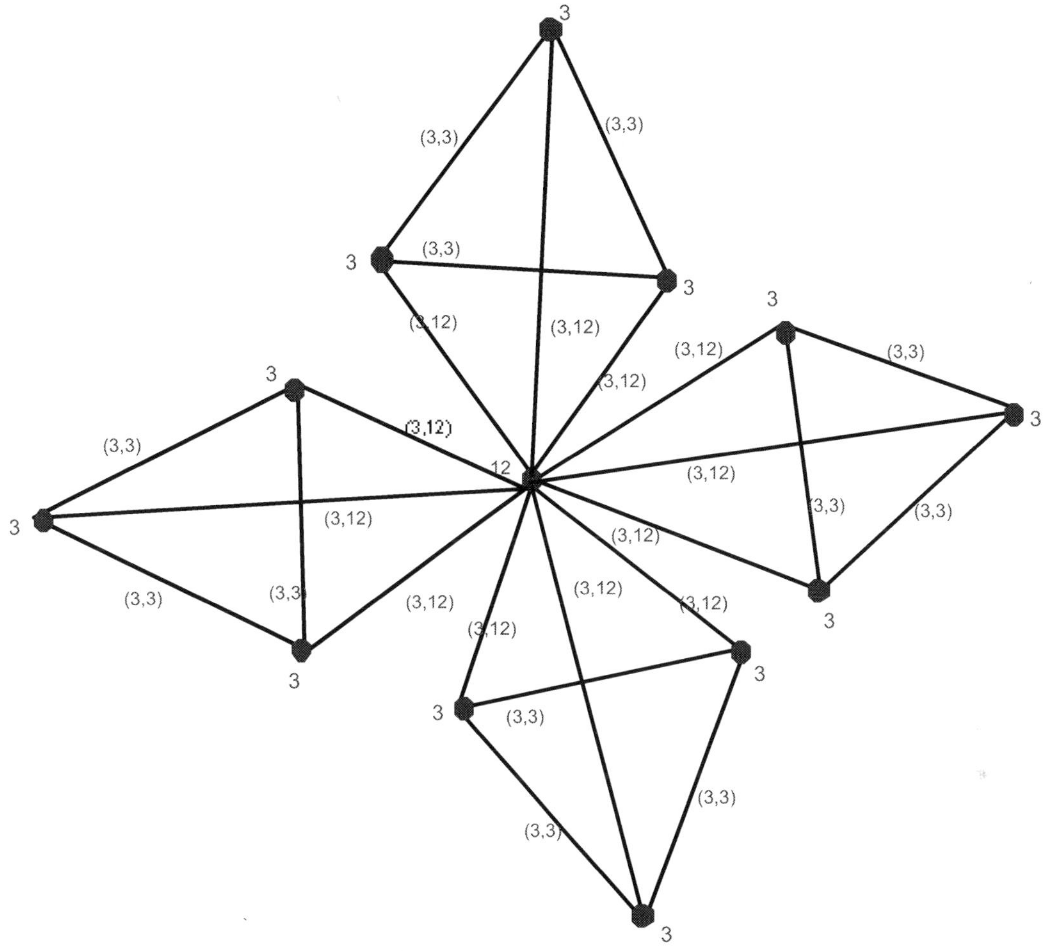

Figure 3.13: A windmill graph W_4^m

From above figure we can easily generalize the vertex and edge partition of W_4^mgraph.

The edge partition of D_4^m graph is given as

Table 27. Edge partition of W_4^m graph.

W_4^m	
$d_u,d_v)$	Number of edges

(2 , 2)	3m
(2 , 3m)	3m

The vertex partition of W_4^m graph is given as

Table 28. Vertex partition of W_4^m graph.

W_4^m	
d_u	Number of vertex
3	3m
3m	1

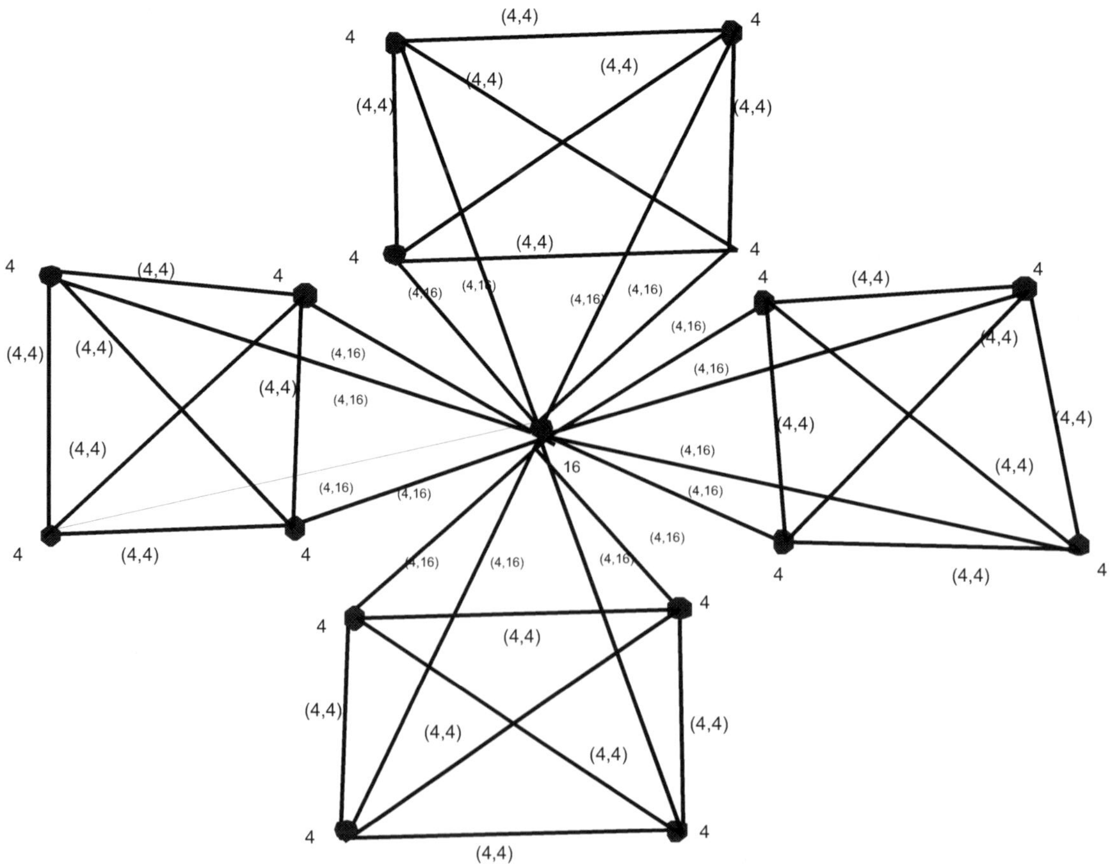

Figure 3.14: A windmill graph W_5^m

From above figure we can easily generalize the vertex and edge partition of W^m_4graph.

The edge partition of D^m_4 graph is given as

Table 29. Edge partition of W^m_5 graph.

w^m_5	
$d_u, d_v)$	Number of edges
(4 , 4)	6m
(4, 4m)	4m

The vertex partition of W^m_5 graph is given as

Table 30. Vertex partition of W^m_5 graph.

W^m_4	
d_u	Number of vertex
4	4m
4m	1

From all of above example now we can easily generalize edge partition of W^m_n windmill graph.

Since K_n is a complete graph and we know that size of complete graph is always $\frac{n(n-1)}{2}$.

So, size of windmill graph is always $\frac{n(n-1)m}{2}$

From all above example of windmill graph have two kind of edges

The edges of degrees $\{(n-1), m(n-1)\}$ are always $m(n-1)$. remaining edges are obtained by subtracting these $(n-1)m$ edges from size of wind mill graph

$$\frac{nm(n-1)}{2} - (n-1)m$$

$$= (\frac{n(n-1)}{2} - n + 1)m$$

$$= m(\frac{n^2 - n - 2n + 2}{2})$$

$$= \frac{(n-1)m(n-2)}{2}$$

From all these observation, now we can easily generalize vertex partition of W_n^m windmill graph.

Table 31. Edge partition of W_n^m graph.

W_n^m	
(d_u, d_v)	No. of edges
((n-1) , m (n-1))	m(n-1)
((n-1) , (n-1))	$\frac{(n-2)m(n-1)}{2}$

From all these observation now we can easily generalize vertex partition of W_n^m windmill graph.

Table 32. Vertex partition of W_n^m graph.

W_n^m	
d_u	No. of vertices
(n-1)	m(n-1)
m(n-1)	1

Wheel graph

Theorem 7.1. Let $G = W_n$ then

$irr(G) = n|n-3|$, $irr_t(G) = \frac{n}{2}|n-3|$,

$$var(G) = \frac{n}{(n+1)^2}(n^2 - 6n + 9).$$

Proof: - We know that $|V(G)| = n + 1.|E(G)| = 2n.$ Vertex and edge partition of this graph is given as

$$V_1(G) = \{v \in V(G): d_G(v) = 3\}$$

$$V_2(G) = \{v \in V(G): d_G(v) = n\}$$

$$E_1(G) = \{e = uv \in E(G): d_G(u) = d_G(v) = 3\}$$

$$E_2(G) = \{e = uv \in E(G): d_G(u) = 5, d_G(v) = 3\}$$

Then,

$$|V_1(G)| = n$$

$$|V_2(G)| = 1$$

$$|E_1(G)| = |E_2(G)| = n$$

$$|E_4(G)| = n - 2$$

Thus, the irregularity irr(G), total irregularity $irr_t(G)$, and variance of Fan graph are

$$irr(G) = \sum_{e \in E(G)} |d_G(u) - d_G(v)| = n|3-3| + n|n-3|$$

$$irr(G) = n|n-3|$$

$$irr_t(G) = \frac{1}{2}\sum_{u,v \in V(G)} |d_G(u) - d_G(v)| = \frac{1}{2}\{n(1)|n-3|\}$$

$$irr_t(G) = \frac{n}{2}|n-3|$$

$$var(G) = \frac{1}{n}\sum_{i=1}^{n} d_i^2 - \frac{1}{n^2}\left(\sum_{i=1}^{n} d_i\right)^2$$

$$var(G) = \frac{1}{(n+1)}\{n3^2 + 1(n^2)\} - \frac{1}{(n+1)^2}\{3(n) + n\}^2$$

$$var(G) = \frac{1}{(n+1)^2}\{(n+1)(9n + n^2) - (4n)^2\}$$

$$var(G) = \frac{n}{(n+1)^2}(n^2 - 6n + 9)$$

Fan graph

Theorem 7.2. Let $G = F_n$ then

$irr(G) = n^2 - 3n + 4$, $irr_t(G) = \frac{1}{2}(n-2)(n+1)$,

$$var(G) = \frac{1}{(n+1)^2}(n^3 - 6n + 15n^2 - 14).$$

Proof: - We know that $|V(G)| = n + 1.|E(G)| = 2n - 1$. Vertex and edge partition of this graph is given as

$V_1(G) = \{v \in V(G): d_G(v) = 2\}$

$V_2(G) = \{v \in V(G): d_G(v) = 3\}$

$V_3(G) = \{v \in V(G): d_G(v) = n\}$

$E_1(G) = \{e = uv \in E(G): d_G(u) = d_G(v) = 3\}$

$E_2(G) = \{e = uv \in E(G): d_G(u) = 3, d_G(v) = 2\}$

$E_3(G) = \{e = uv \in E(G): d_G(u) = n, d_G(v) = 2\}$

$E_4(G) = \{e = uv \in E(G): d_G(u) = n, d_G(v) = 3\}$

Then,

$|V_1(G)| = 2, |V_2(G)| = n - 2, |V_3(G)| = 1$

$|E_1(G)| = n - 3,\ |E_2(G)| = |E_3(G)| = 2,\ |E_4(G)| = n - 2$

Thus, the irregularity irr(G), total irregularity $irr_t(G)$, and variance of Fan graph are

$$irr(G) = \sum_{e \in E(G)} |d_G(u) - d_G(v)| = (n-3)|3-3| + 2|3-2| + 2|n-2| + (n-2)|n-3|$$

$$irr(G) = 2 + 2|n-2| + (n-2)|n-3| = n^2 - 3n + 4$$

$$irr_t(G) = \frac{1}{2}\sum_{u,v \in V(G)} |d_G(u) - d_G(v)| = \frac{1}{2}\{2(n-2)|3-2| + 2(1)|n-2| + 2(n-2)|n-3|\}$$

$$irr_t(G) = \frac{1}{2}(n-2)(n+1)$$

$$var(G) = \frac{1}{n}\sum_{i=1}^{n} d_i^2 - \frac{1}{n^2}\left(\sum_{i=1}^{n} d_i\right)^2$$

$$var(G) = \frac{1}{(n+1)}\{2(2^2) + (n-2)3^2 + 1(n^2)\} - \frac{1}{(n+1)^2}\{2(2) + 3(n-2) + n\}^2$$

$$var(G) = \frac{1}{(n+1)^2}\{(n+1)(9n + n^2 - 10) - (4n-2)^2\}$$

$$var(G) = \frac{1}{(n+1)^2}(n^3 - 6n^2 + 15n - 14)$$

Helm graph

Theorem 7.3. Let $G = H_n$ then

$$irr(G) = \begin{cases} n^2 - n & when\ n \geq 4 \\ 7n - n^2 & when\ n \leq 4 \end{cases}$$

$$irr_t(G) = \begin{cases} \dfrac{5n(n-1)}{2} & n \geq 4 \\ \dfrac{n(n+3)}{2} & n < 4 \end{cases},$$

$$var(G) = \frac{n}{(2n+1)^2}(2n^2 - n + 17).$$

Proof:- We know that $|V(G)| = 2n + 1. |E(G)| = 3n.$ Vertex and edge partition of this graph is given as

$V_1 = \{v \in V: d_G(v) = 1\}$

$V_2 = \{v \in V: d_G(v) = 4\}$

$V_3 = \{v \in V: d_G(v) = n\}$

$E_1 = \{e = uv \in E: d_G(u) = d_G(v) = 4\}$

$E_2 = \{e = uv \in E: d_G(u) = 1, d_G(v) = 4\}$

$E_3 = \{e = uv \in E(G): d_G(u) = n, d_G(v) = 4\}$

Then,

$|V_1| = |V_2| = n, |V_3| = 1$

$|E_1(G)| = |E_2(G)| = |E_3(G)| = n$

Thus, the irregularity irr(G), total irregularity $irr_t(G)$, and variance of Helm graph are

$$irr(G) = \sum_{e \in E(G)} |d_G(u) - d_G(v)| = n|4-4| + n|4-1| + n|n-4|$$

When $n \geq 4$

$$irr(G) = 3n + n^2 - 4n = n^2 - n$$

When $n \leq 4$

$$irr(G) = 3n + 4n - n^2 = 7n - n^2$$

$$irr_t(G) = \frac{1}{2} \sum_{u,v \in V(G)} |d_G(u) - d_G(v)| = \frac{1}{2}\{n(n)|4-1| + n(1)|n-1| + n(1)|n-4|\}$$

When $n \geq 4$

$$irr_t(G) = \frac{n}{2}\{3n + n - 1 + n - 4\} = \frac{5n}{2}(n-1)$$

When $n \leq 4$

$$irr_t(G) = \frac{n}{2}(3n + n - 1 + 4 - n) = \frac{3n}{2}(n+3)$$

$$var(G) = \frac{1}{n}\sum_{i=1}^{n} d_i^2 - \frac{1}{n^2}\left(\sum_{i=1}^{n} d_i\right)^2$$

$$var(G) = \frac{1}{(2n+1)}\{n(1) + n(4^2) + 1(n^2)\} - \frac{1}{(2n+1)^2}\{n(1) + n(4) + n\}^2$$

$$var(G) = \frac{1}{(2n+1)^2}\{(2n+1)(17n + n^2) - (6n)^2\}$$

$$var(G) = \frac{1}{(2n+1)^2}(2n^3 + n^2 + 34n^2 + 17n - 36n^2)$$

$$var(G) = \frac{n}{(2n+1)^2}(2n^2 - n + 17)$$

Web graph

Theorem 7.4. Let $W = Wb_n$ then

$$irr(W) = \begin{cases} n^2 - n & when\, n \geq 4 \\ 7n - n^2 & when\, n \leq 4 \end{cases}$$

$$irr_t(W) = \begin{cases} \frac{9n(n-1)}{2} & n \geq 4 \\ \frac{n(5n+7)}{2} & n < 4 \end{cases},$$

$$var(G) = \frac{3n}{(3n+1)^2}(n^2 + 11).$$

Proof:- We know that $|V(W)| = 3n + 1. |E(W)| = 5n.$ Vertex and edge partition of this graph is given as

$V_1(W) = \{v \in V(W): d_G(v) = 1\}$

$V_2(W) = \{v \in V(W): d_G(v) = 4\}$

$V_3(W) = \{v \in V(W): d_G(v) = n\}$

$E_1(W) = \{e = uv \in E(W): d_G(u) = d_G(v) = 4\}$

$E_2(W) = \{e = uv \in E(W): d_G(u) = 1, d_G(v) = 4\}$

$E_3(W) = \{e = uv \in E(W): d_G(u) = n, d_G(v) = 4\}$

Then,

$|V_1(W)| = n$

$|V_2(W)| = 2n$

$|V_3(W)| = 1$

$|E_1(W)| = 3n$

$|E_2(W)| = |E_3(W)| = n$

Thus, the irregularity irr(W), total irregularity $irr_t(W)$, and variance of Web graph are

$$irr(W) = \sum_{e \in E(G)} |d_G(u) - d_G(v)| = 3n|4 - 4| + n|4 - 1| + n|n - 4|$$

When $n \geq 4$

$$irr(W) = 3n + n^2 - 4n = n^2 - n$$

When $n \leq 4$

$$irr(W) = 3n + 4n - n^2 = 7n - n^2$$

$$irr_t(W) = \frac{1}{2} \sum_{u,v \in V(G)} |d_G(u) - d_G(v)| = \frac{1}{2}\{n(2n)|4 - 1| + n(1)|n - 1| + 2n(1)|n - 4|\}$$

When $n \geq 4$

$$irr_t(W) = \frac{n}{2}\{6n + n - 1 + 2n - 8\} = \frac{9n}{2}(n - 1)$$

When $n \leq 4$

$$irr_t(W) = \frac{n}{2}(6n + n - 1 + 8 - 2n) = \frac{n}{2}(5n + 7)$$

$$var(W) = \frac{1}{n}\sum_{i=1}^{n} d_i^2 - \frac{1}{n^2}\left(\sum_{i=1}^{n} d_i\right)^2$$

$$var(W) = \frac{1}{(3n + 1)}\{n(1) + 2n(4^2) + 1(n^2)\} - \frac{1}{(3n + 1)^2}\{n(1) + 2n(4) + n\}^2$$

$$var(W) = \frac{1}{(3n + 1)^2}\{(3n + 1)(33n + n^2) - (10n)^2\}$$

$$var(W) = \frac{1}{(3n + 1)^2}(99n^2 + 33n + 3n^3 + n^2 - 100n^2)$$

$$var(W) = \frac{3n}{(3n + 1)^2}(n^2 + 11)$$

Flower graph

Theorem 7.5. Let $G = Fl_n$ then

$irr(G) = 4n(n-1)$, $irr_t(G) = \frac{6}{2}(n-1)$,

$$var(G) = \frac{n}{(2n+1)^2}(8n^2 - 20n + 20).$$

Proof:- We know that $|V(G)| = 2n + 1.|E(G)| = 4n.$ Vertex and edge partition of this graph is given as

$V_1 = \{v \in V(G): d_G(v) = 2\}$

$V_2 = \{v \in V(G): d_G(v) = 4\}$

$V_3 = \{v \in V(G): d_G(v) = 2n\}$

$E_1 = \{e = uv \in E(G): d_G(u) = 4,\ d_G(v) = 2\}$

$E_2 = \{e = uv \in E(G): d_G(u) = 2n, d_G(v) = 2\}$

$E_3 = \{e = uv \in E(G): d_G(u) = 4, d_G(v) = 4\}$

$E_4 = \{e = uv \in E(G): d_G(u) = 2n, d_G(v) = 4\}$

Then,

$|V_1| = |V_2| = n$

$|V_3| = 1$

$|E_1| = |E_2| = |E_3| = |E_4| = n$

Thus, the irregularity irr(G), total irregularity $irr_t(G)$, and variance of Fan graph are

$$irr(G) = \sum_{e \in E(G)} |d_G(u) - d_G(v)| = n|4-2| + n|2n-2| + n|4-4| + n|2n-4|$$

$$irr(G) = 2n + 2n(n-1) + 2n(n-2) = 4n(n-1)$$

$$irr_t(G) = \frac{1}{2}\sum_{u,v \in V(G)} |d_G(u) - d_G(v)| = \frac{1}{2}\{n(n)|4-2| + n(1)|2n-2| + n|4n-4|\}$$

$$irr_t(G) = \frac{6n}{2}(n-1).$$

$$var(G) = \frac{1}{n}\sum_{i=1}^{n} d_i^{\,2} - \frac{1}{n^2}\left(\sum_{i=1}^{n} d_i\right)^2$$

$$var(G) = \frac{1}{(2n+1)}\{n(2^2) + n4^2 + 1(2n)^2\} - \frac{1}{(n+1)^2}\{2(n) + 4(n) + 2n\}^2$$

$$var(G) = \frac{1}{(2n+1)^2}\{(2n+1)(20n + 4n^2) - (8n)^2\}$$

$$var(G) = \frac{n}{(2n+1)^2}(8n^2 - 20n + 20)$$

Sunflower graph

Theorem 7.6. Let $G = SFl_n$ then

$irr(G) = 9n^2 - 5n$, $irr_t(G) = \frac{n}{2}(15n - 7)$,

$$var(G) = \frac{n}{(3n+1)^2}(27n^2 - 28n + 21).$$

Proof:- We know that $|V(G)| = 3n + 1.|E(G)| = 5n.$ Vertex and edge partition of this graph is given as

$V_1 = \{v \in V(G): d_G(v) = 1\}$

$V_2 = \{v \in V(G): d_G(v) = 2\}$

$V_3 = \{v \in V(G): d_G(v) = 4\}$

$V_4 = \{v \in V(G): d_G(v) = 3n\}$

$E_1 = \{e = uv \in E(G): d_G(u) = d_G(v) = 4\}$

$E_2 = \{e = uv \in E(G): d_G(u) = 3n, d_G(v) = 2\}$

$E_3 = \{e = uv \in E(G): d_G(u) = 3n, d_G(v) = 1\}$

$E_4 = \{e = uv \in E(G): d_G(u) = 3n, d_G(v) = 4\}$

$E_5 = \{e = uv \in E(G): d_G(u) = 2, d_G(v) = 4\}$

Then,

$|V_1| = |V_2| = |V_3| = n$

$|V_4| = 1$

$|V_4| = n$

$|E_1| = |E_2| = |E_3| = |E_4| = |E_5| = n$

Thus, the irregularity irr(G), total irregularity $irr_t(G)$, and variance of Fan graph are

$$irr(G) = \sum_{e \in E(G)} |d_G(u) - d_G(v)| = n|4-4| + n|3n-2| + n|3n-1| + n|3n-4| + n|4-2|$$

$$irr(G) = n(3n-2) + n(3n-1) + n(3n-4) + 2n = 9n^2 - 5n$$

$$irr_t(G) = \frac{1}{2} \sum_{u,v \in V(G)} |d_G(u) - d_G(v)|$$

$$irr_t(G) = \frac{1}{2}\{n^2|2-1| + n^2|4-1| + n|3n-1| + n^2|4-2| + n|3n-2| + n|3n-4|\}$$

$$irr_t(G) = \frac{1}{2}\{n^2 + 3n^2 + 3n^2 - n + 2n^2 + 3n^2 - 2n + 3n^2 - 4n\}$$

$$irr_t(G) = \frac{n}{2}(15n-7)$$

$$var(G) = \frac{1}{n}\sum_{i=1}^{n} d_i^2 - \frac{1}{n^2}\left(\sum_{i=1}^{n} d_i\right)^2$$

$$var(G) = \frac{1}{(3n+1)}\{n(1^2) + n(2^2) + n(4^2) + 1(3n)^2\} - \frac{1}{(3n+1)^2}\{n(1) + n(2) + n(4) + 3n\}^2$$

$$var(G) = \frac{1}{(3n+1)^2}\{(3n+1)(21n+9n^2) - (10n)^2\}$$

$$var(G) = \frac{n}{(n+1)^2}(27n^2 - 28n + 21)$$

Gear graph

Theorem 7.7. Let $G = G_n$ then

$irr(G) = n^2 - n$, $irr_t(G) = \frac{n}{2}(3n - 5)$,

$$var(G) = \frac{n}{(2n + 1)^2}(2n^2 - 9n + 13).$$

Proof:- We know that $|V(G)| = 2n + 1.|E(G)| = 3n.$ Vertex and edge partition of this graph is given as

$$V_1 = \{v \in V(G): d_G(v) = 3\}$$

$$V_2 = \{v \in V(G): d_G(v) = 2\}$$

$$V_3 = \{v \in V(G): d_G(v) = n\}$$

$$E_1 = \{e = uv \in E(G): d_G(u) = 3, d_G(v) = 2\}$$

$$E_2 = \{e = uv \in E(G): d_G(u) = n, d_G(v) = 3\}$$

Then,

$$|V_1| = |V_2| = n$$

$$|V_3| = 1$$

$$|E_1| = 2n$$

$$|E_2| = n$$

Thus, the irregularity irr(G), total irregularity $irr_t(G)$, and variance of Fan graph are

$$irr(G) = \sum_{e \in E(G)} |d_G(u) - d_G(v)| = 2n|3 - 2| + n|n - 3|$$

$$irr(G) = 2n + n^2 - 3n = n^2 - n$$

$$irr_t(G) = \frac{1}{2}\sum_{u,v \in V(G)} |d_G(u) - d_G(v)| = \frac{1}{2}\{n^2|3 - 2| + n|n - 3| + n|n - 2|\}$$

$$irr_t(G) = \frac{1}{2}\{n^2 + n^2 - 3n + n^2 - 2n\}$$

$$irr_t(G) = \frac{n}{2}(3n - 5)$$

$$var(G) = \frac{1}{n}\sum_{i=1}^{n} d_i^{\,2} - \frac{1}{n^2}\left(\sum_{i=1}^{n} d_i\right)^2$$

$$var(G) = \frac{1}{(2n+1)}\{n(3^2) + n(2^2) + 1(n^2)\} - \frac{1}{(2n+1)^2}\{n(3) + n(1) + 1(n)\}^2$$

$$var(G) = \frac{1}{(2n+1)^2}\{(2n+1)(13n+n^2) - (6n)^2\}$$

$$var(G) = \frac{n}{(n+1)^2}(2n^2 - 9n + 13)$$

Closed Helm graph

Theorem 7.8. Let $G = CH_n$ then

$$irr(G) = \begin{cases} n^2 - 3n & n \geq 4 \\ 5n - n^2 & n \leq 4 \end{cases},$$

$$irr_t(G) = \begin{cases} \frac{n}{2}(3n-7) & n \geq 4 \\ \frac{n}{2}(7-n) & n < 4 \end{cases},$$

$$var(G) = \frac{n}{(2n+1)^2}(2n^2 - 13n + 25).$$

Proof:- We know that $|V(G)| = 2n + 1.|E(G)| = 4n.$ Vertex and edge partition of this graph is given as

$V_1 = \{v \in V(G) : d_G(v) = 3\}$

$V_2 = \{v \in V(G) : d_G(v) = 4\}$

$V_3 = \{v \in V(G) : d_G(v) = n\}$

$E_1 = \{e = uv \in E(G) : d_G(u) = 3, d_G(v) = 4\}$

$E_2 = \{e = uv \in E(G) : d_G(u) = 3, d_G(v) = 3\}$

$E_3 = \{e = uv \in E(G) : d_G(u) = 4\,,\, d_G(v) = 4\}$

$E_4 = \{e = uv \in E(G) : d_G(u) = n, d_G(v) = 4\}$

Then,

$|V_1| = |V_2| = n,\ |V_3| = 1$

$|E_1| = |E_2| = |E_3| = |E_4| = n$

Thus, the irregularity irr(G), total irregularity $irr_t(G)$, and variance of Fan graph are

$$irr(G) = \sum_{e \in E(G)} |d_G(u) - d_G(v)| = n|4-3| + n|3-3| + n|4-4| + n|n-4|$$

When $n \geq 4$

$$irr(G) = n + n^2 - 4n = n^2 - 3n$$

When $n \leq 4$

$$irr(G) = n + 4n - n^2 = 5n - n^2$$

$$irr_t(G) = \frac{1}{2} \sum_{u,v \in V(G)} |d_G(u) - d_G(v)| = \frac{1}{2}\{n(n)|4-3| + n(1)|n-3| + n(1)|n-4|\}$$

When $n \geq 4$

$$irr_t(G) = \frac{n}{2}(n^2 + n^2 - 3n + 3n^2 - 4n) = \frac{n}{2}(3n - 7)$$

When $n < 4$

$$irr_t(G) = \frac{n}{2}(n^2 + 3n - n^2 + 4n - n^2) = \frac{n}{2}(7 - n)$$

$$var(G) = \frac{1}{n}\sum_{i=1}^{n} d_i^{\,2} - \frac{1}{n^2}\left(\sum_{i=1}^{n} d_i\right)^2$$

$$var(G) = \frac{1}{(2n+1)}\{n(3^2) + n(4^2) + 1(n^2)\} - \frac{1}{(2n+1)^2}\{n(3) + n(4) + n(1)\}^2$$

$$var(G) = \frac{1}{(2n+1)^2}\{(2n+1)(25n + n^2) - (8n)^2\}$$

$$var(G) = \frac{1}{(2n+1)^2}(50n^2 + 25n + 2n^3 + n^2 - 64n^2)$$

$$var(G) = \frac{n}{(2n+1)^2}(2n^2 - 13n + 25)$$

Dutch Windmill graph

Theorem 7.9. Let $G = D_n^m$ then

$$irr(G) = 4m(m-1)\ , \quad irr_t(G) = \frac{n}{2}(3n-5)\ ,$$

$$var(G) = \frac{4m}{(mn-m+1)^2}(n(1-2m) + m^2(n-1) + 2m - 1).$$

Proof:- We know that $|V(G)| = mn - (m-1).|E(G)| = mn.$ Vertex and edge partition of this graph is given as

$$V_1(G) = \{v \in V(G) \colon d_G(v) = 2\}$$

$$V_2(G) = \{v \in V(G) \colon d_G(v) = 2m\}$$

$$E_1(G) = \{e = uv \in E(G) \colon d_G(u) = 2, d_G(v) = 2\}$$

$$E_2(G) = \{e = uv \in E(G) \colon d_G(u) = 2m\ , d_G(v) = 2\}$$

Then,

$$|V_1(G)| = m(n-1)$$

$$|V_2(G)| = 1$$

$$|E_1(G)| = m(n-2)$$

$$|E_2(G)| = 2m$$

Thus, the irregularity irr(G), total irregularity $irr_t(G)$, and variance of Fan graph are

$$irr(G) = \sum_{e \in E(G)} |d_G(u) - d_G(v)| = m(n-2)|2-2| + 2m|2m-2|$$

$$irr(G) = 4m(m-1)$$

$$irr_t(G) = \frac{1}{2}\sum_{u,v \in V(G)} |d_G(u) - d_G(v)| = \frac{1}{2}\{m(n-1)(1)|2m-2|\}$$

$$irr_t(G) = \frac{2}{2}\{m(m-1)(n-1)\}$$

$$irr_t(G) = m(m-1)(n-1)$$

$$var(G) = \frac{1}{n}\sum_{i=1}^{n} d_i^{\,2} - \frac{1}{n^2}\left(\sum_{i=1}^{n} d_i\right)^2$$

$$var(G) = \frac{\{m(n-1)(2^2) + (1)(2m)^2\}}{(mn-m+1)} - \frac{\{m(n-1)(2) + 2m\}^2}{(mn-m+1)^2}$$

$$var(G) = \frac{1}{(mn-m+1)^2}\{(mn-m+1)(4mn-4m+4m^2) - (4mn)^2\}$$

$$var(G) = \frac{n}{(mn-m+1)^2}(4mn - 8m^2n + 4m^3n - 4m^3 + 8m^2 - 4m)$$

$$var(G) = \frac{4m(n(1-2m) + m^2(n-1) + 2m - 1)}{(mn-m+1)^2}.$$

Windmill graph

Theorem 7.10. Let $G = W_{m,n}$ then

$$irr(G) = m(n-1)|(n-1) - m(n-1)|\ ,$$

$$irr_t(G) = \frac{1}{2}\{m(n-1)|(n-1) - m(n-1)|\}\ ,$$

$$var(G) = \frac{1}{(A)^2}[(A)\{m(n-1)^3 + m^2(n-1)^2\} - \{m(n-1)^2 + m(n-1)\}^2]$$

Proof:- $|V(G)| = mn - (m-1) = A.$

$|E(G)| = \frac{mn(n-1)}{2}.$ Vertex and edge partition of this graph is given as

$$V_1 = \{v \in V(G) \colon d_G(v) = n-1\}$$

$$V_2 = \{v \in V(G) \colon d_G(v) = 1\}$$

$$E_1(G) = \{e = uv \in E(G) \colon d_G(u) = n-1, d_G(v) = m(n-1)\}$$

$$E_2(G) = \{e = uv \in E(G) \colon d_G(u) = d_G(v) = n-1\}$$

Then,

$$|V_1(G)| = m(n-1)$$

$$|V_2(G)| = 1$$

$$|E_1(G)| = m(n-2)$$

$$|E_2(G)| = \frac{m(n-2)(n-1)}{2}$$

Thus, the irregularity irr(G), total irregularity $irr_t(G)$, and variance of Fan graph are

$$irr(G) = \sum_{e \in E(G)} |d_G(u) - d_G(v)|$$

$$irr(G) = m(n-1)|(n-1) - m(n-1)| + \frac{(n-2)(n-1)m}{2}|(n-1) - (n-1)|$$

$$irr(G) = m(n-1)|(n-1) - m(n-1)|$$

$$irr_t(G) = \frac{1}{2}\sum_{u,v \in V(G)} |d_G(u) - d_G(v)| = \frac{1}{2}\{m(n-1)(1)|(n-1) - (n-1)m|\}$$

$$var(G) = \frac{1}{n}\sum_{i=1}^{n} d_i^{\,2} - \frac{1}{n^2}\left(\sum_{i=1}^{n} d_i\right)^2$$

$$var(G) = \frac{1}{(A)}\{m(n-1)(n-1)^2 + m^2(n-1)^2\} - \frac{1}{(A)^2}\{m(n-1)^2 + m(n-1)\}^2$$

$$var(G) = \frac{1}{(A)^2}[(A)\{m(n-1)^3 + m^2(n-1)^2\} - \{m(n-1)^2 + m(n-1)\}^2]$$

Bibliography

1. Agnarsson, G. and Greenlaw, R., 2006. *Graph theory: Modeling, applications, and algorithms*. Prentice-Hall, Inc..
2. Biggs, N., Lloyd, E.K. and Wilson, R.J., 1986. *Graph Theory, 1736-1936*. Oxford University Press.
3. Bondy, J.A. and Murty, U.S.R., 1976. *Graph theory with applications* (Vol. 290). London: Macmillan.
4. Erdos, P., 2011. PROBLEMS IN GRAPH THEORY". *Combinatorics 79. Part II, 9,* pp.37-42.
5. Gross, J.L., Yellen, J. and Anderson, M., 2018. *Graph theory and its applications*. Chapman and Hall/CRC.
6. Hamada, T. and Yoshimura, I., 1976. Traversability and connectivity of the middle graph of a graph. *Discrete Mathematics, 14*(3), pp.247-255.
7. Mader, W., 1978. A reduction method for edge-connectivity in graphs. In *Annals of Discrete Mathematics* (Vol. 3, pp. 145-164). Elsevier.
8. Marcus, D.A., 2020. *Graph theory* (Vol. 53). American Mathematical Soc..
9. Phillips, J.D., Schwanghart, W. and Heckmann, T., 2015. Graph theory in the geosciences. *Earth-Science Reviews, 143,* pp.147-160.
10. Rahman, M.S., 2017. *Basic graph theory* (Vol. 9). India: Springer.
11. Vasudev, C., 2006. *Graph theory with applications*. New Age International.
12. Wallis, W.D., 2007. *A beginner's guide to graph theory*. Springer Science & Business Media.
13. Xu, J., 2003. *Theory and application of graphs* (Vol. 10). Springer Science & Business Media.
14. Zhang, P. and Chartrand, G., 2006. *Introduction to graph theory*. Tata McGraw-Hill.

Made in United States
Orlando, FL
05 November 2026

90244041R00072